Perspectives on Development in the Middle East and North Africa (MENA) Region

Series Editor

Almas Heshmati, Jönköping University, Jönköping, Sweden

This book series publishes monographs and edited volumes devoted to studies on the political, economic and social developments of the Middle East and North Africa (MENA). Volumes cover in-depth analyses of individual countries, regions, cases and comparative studies, and they include both a specific and a general focus on the latest advances of the various aspects of development. It provides a platform for researchers globally to carry out rigorous economic, social and political analyses, to promote, share, and discuss current quantitative and analytical work on issues, findings and perspectives in various areas of economics and development of the MENA region. Perspectives on Development in the Middle East and North Africa (MENA) Region allows for a deeper appreciation of the various past, present, and future issues around MENA's development with high quality, peer reviewed contributions. The topics may include, but not limited to: economics and business, natural resources, governance, politics, security and international relations, gender, culture, religion and society, economics and social development, reconstruction, and Jewish, Islamic, Arab, Iranian, Israeli, Kurdish and Turkish studies. Volumes published in the series will be important reading offering an original approach along theoretical lines supported empirically for researchers and students, as well as consultants and policy makers, interested in the development of the MENA region.

More information about this series at http://www.springer.com/series/13870

Masoomeh Rashidghalam
Editor

The Economics of Agriculture and Natural Resources

The Case of Iran

Springer

Editor
Masoomeh Rashidghalam
Department of Agricultural Economics
University of Tabriz
Tabriz, Iran

ISSN 2520-1239 ISSN 2520-1247 (electronic)
Perspectives on Development in the Middle East and North Africa (MENA) Region
ISBN 978-981-15-5249-6 ISBN 978-981-15-5250-2 (eBook)
https://doi.org/10.1007/978-981-15-5250-2

This Springer imprint is published by the registered company Springer Nature Singapore Pte Ltd.
The registered company address is: 152 Beach Road, #21-01/04 Gateway East, Singapore 189721, Singapore

To my parents, Mohammad and Zari

Contents

Introduction to the Economics of Agriculture and Natural Resources 1
Masoomeh Rashidghalam

Agricultural Resilience and Sustainability

Quantitative Sustainability Assessment Applied to Dairy Farms 11
Leila Hassani, Mahmoud Daneshvar kakhki, Mahmoud Sabuhi sabouni, and Peter Fantke

A Framework for Economic Resilience Assessment of Agricultural Production Systems 21
Leila Hassani, Mahmoud Daneshvar kakhki, Mahmoud Sabuhi sabouni, and Peter Fantke

Empirical Evaluation of Agricultural Sustainability Using Entropy and FAHP Methods 31
Marziyeh Manafi Mollayosefi, Babollah Hayati, Esmaeil Pishbahar, and Javad Nematian

Agricultural Producers and Consumers

Weather Risk Management: The Application of Vine Copula Approach 49
Sasan Torabi, Arash Dourandish, Mahmoud Daneshvar Kakhki, Ali Kianirad, and Hosein Mohammadi

An Investigation on Dependency Structure Between Temperature-Humidity Index (THI) and Milk Yield 73
Afsaneh Nikoukar and Sasan Torabi

Adoption of IPM by Farmland Owners and Non-owners: Application of Endogenous Switching Copula Approach 91
Sahar Abedi, Pariya Bagheri, and Esmaeil Pishbahar

Estimating Economic Value and Compensation Surplus of Animal Species in Arasbaran Forests Using Choice Experiment Method 109
Maryam Haghjou, Babollah Hayati, Esmaeil Pishbahar, and Morteza Molaei

Factors Affecting Consumers' Awareness of Pesticides-Free Fruits and Vegetables 125
Maryam Haghjou, Babollah Hayati, and Esmaeil Pishbahar

Energy Use in Agriculture

The Relationship Between Economic Growth, Energy Consumption, and CO_2 Emissions 143
Parisa Pakrooh and Esmaeil Pishbahar

Oil Price Volatility and Food Price Linkage: Evidence of Dutch Disease in Iran's Agricultural Sector 171
Zahra Rasouli, Mohammad Ghahremanzadeh, and Masoomeh Rashidghalam

Environmental Efficiency in Agricultural Sector 183
Pariya Bagheri, Sahar Abedi, and Farid Bagheri Sarajug

Author Index 203

Subject Index 205

About the Editor

Masoomeh Rashidghalam is a visiting researcher at Centre for Entrepreneurship and Spatial Economics (CEnSE), Jönköping International Business School (JIBS), Jönköping, Sweden. She did her B.Sc. and Ph.D. in the Department of Agricultural Economics at the University of Tabriz and holds an M.Sc. from Tarbiat Modares University. Dr. Rashidghalam's areas of expertise are Agricultural Production Economics, Productivity and efficiency, Well-Being, and Urbanization. She has a wide range of teaching experience in Econometrics, Agricultural Production Economics, and Microeconomics. She has written two books: *Measurement and Analysis of Performance of Industrial Crop Production: The Case of Iran's Cotton and Sugar Beet Production*, 2018, published by Springer. *Sustainable Agriculture and Agribusiness in Iran*, 2019, published by Springer. She has publications in the *Journal of Productivity Analysis*.

Abbreviations

AHP	Analytic hierarchy process
AIC	Akaike information criterion
AMH	Ali-Mikhail-Hagh
ANA	Attribute non-attendance
ARDL	Autoregressive distributed lag
ARIMA	Autoregressive integrated moving average
ATE	Average treatment effect
BIC	Bayesian information criterion
CDF	Cumulative distribution function
CE	Choice experiment
CI	Composite indicator
CRI	Cumulative rainfall index
C-vine	Canonical vine copula
DEA	Data Envelopment Analysis
DM	Decision making
D-vine	Drawable vine copula
EAA	Endogenous attribute attendance
EKC	Environmental Kuznets Curve
ERI	Economic resilience indicator
FAHP	Fuzzy analytic hierarchy process
FAO	Food and Agriculture Organization
FDA	Food and Drug Administration
FGM	Farlie-Gumbel-Morgenstern
FIML	Full information maximum likelihood
GARCH	Generalized autoregressive conditional heteroskedasticity
GDP	Gross domestic product
GHG	Greenhouse gas
GMM	Generalized method of moments
IPCC	Intergovernmental panel on climate change
IPM	Integrated Pest Management

MCDM	Multi-criteria decision making
MCMC	Markov chain Monte Carlo
MGA	Modeling to generate alternatives
MNL	Multi nominal logit
NRC	Natural Resource Research
OLS	Ordinary least squares
PAM	Partitioning around medoids
PC	Principal components analysis
PCC	Pair copula constructions
PFFV	Pesticide-free fruits and vegetables
PGARCH	Periodic generalized autoregressive conditional heteroskedasticity
PSE	Producer support estimates
RDI	Rural Diversity Index
RH	Relative humidity
RIMA	Resilience index measurement and analysis
RUM	Random utility modeling
R-vine	Regular vine copula
SCC	Somatic Cell Count
SCM	Supply chain management
SI	Sustainability indicator
SP	Stated preferences
TFN	Triangular fuzzy numbers
THI	Temperature-Humidity Index
TOPSIS	Technique for order of preference by similarity to ideal solution
TSP	Traveling salesman problem
TVP-VAR	Time-varying parameter vector autoregression
VAR	Vector autoregressive model
VECM	Vector error correction model
WTP	Willingness to pay

List of Figures

Empirical Evaluation of Agricultural Sustainability Using Entropy and FAHP Methods

Fig. 1 Hierarchical levels. 37

Weather Risk Management: The Application of Vine Copula Approach

Fig. 1 Example of a 5- dimensional C-vine copula's trees 55
Fig. 2 Example of a 5- dimensional D-vine copula's trees 55
Fig. 3 Representation of C-vine tree structure for apple yield and weather variables 61
Fig. 4 Representation of D-vine tree structure for apple yield and weather variables 63

An Investigation on Dependency Structure Between Temperature-Humidity Index (THI) and Milk Yield

Fig. 1 milk yield and THI during 2012–16 81
Fig. 2 The copula's contour plat 85

Adoption of IPM by Farmland Owners and Non-owners: Application of Endogenous Switching Copula Approach

Fig. 1 Integrated pest management components. *Source* Pishbahar et al. (2019) 93

Estimating Economic Value and Compensation Surplus of Animal Species in Arasbaran Forests Using Choice Experiment Method

Fig. 1 The characteristics and their levels of selected animal species of Arasbaran forests 115

Fig. 2 An example of the selection cards to estimate the value three selected animal species' of Arasbaran forests 116

The Relationship Between Economic Growth, Energy Consumption, and CO_2 Emissions

Fig. 1 Kuznets' curve. *Source* Dinda (2004) 146
Fig. 2 Results of TVP-VAR model with stochastic volatility for simulated data. The first row (autocorrelation status of samples), the second row (path of the samples), the third row (pre-density of the samples) 159
Fig. 3 Stochastic volatility of GDP 160
Fig. 4 Stochastic volatility of CO_2. 160
Fig. 5 Stochastic volatility of oil 161
Fig. 6 Stochastic volatility of gas 161
Fig. 7 Stochastic volatility of coal 161
Fig. 8 Stochastic volatility of electricity 162
Fig. 9 Response of CO_2 to GDP shocks 162
Fig. 10 **a** Response of CO_2 to oil shocks. **b** Response of GDP to oil shocks. **c** Response of oil to GDP shocks 163
Fig. 11 **a** Response of GDP to gas shocks. **b** Response of gas to GDP shocks. **c** Response of CO_2 to gas shocks 164
Fig. 12 **a** Response of GDP to coal shocks. **b** Response of coal to GDP shocks. **c** Response of CO_2 to coal shocks 165
Fig. 13 **a** Response of GDP to electricity shocks. **b** Response of electricity to GDP shocks. **c** Response of CO_2 to electricity shocks 166

Oil Price Volatility and Food Price Linkage: Evidence of Dutch Disease in Iran's Agricultural Sector

Fig. 1 Graph of production, subsidy, loan, and imports of Iran agricultural sector **a** production **b** subsidy **c** loan **d** import 172
Fig. 2 Graphs of food price index and FINF in Iran, 2002–2012 176
Fig. 3 Plots of residuals (ε) and conditional volatility (h) estimated from MA(6)/PGARCH(1,1) model 179

Environmental Efficiency in Agricultural Sector

Fig. 1 Distance matrix of provinces in terms of carbon dioxide emissions caused by agricultural sector 193
Fig. 2 The silhouette graph to determine the optimal number of clusters 195
Fig. 3 Provinces clustering based on the carbon dioxide emission using the k-medoids and the silhouette coefficient 196

List of Tables

Quantitative Sustainability Assessment Applied to Dairy Farms

Table 1 Energy content of inputs in dairy farms 14
Table 2 Formulas for calculating equivalent energy inputs of dairy farms. 15
Table 3 Result of equivalent energy indicators of dairy farms. 16
Table 4 Determine sustainability indicator Levels of dairy farms in this study 17

A Framework for Economic Resilience Assessment of Agricultural Production Systems

Table 1 Economic resilience indicator criteria in dairy farms 24
Table 2 Result of quality factors and profitability indicators of Dairy Farms 26
Table 3 Determined economic resilience indicator levels of dairy farms in this study 27

Empirical Evaluation of Agricultural Sustainability Using Entropy and FAHP Methods

Table 1 List of indicators for evaluation of agricultural sustainability. 38
Table 2 Descriptive statistics of agricultural sustainability indicators. . . . 39
Table 3 The relative weights and ranks of agricultural sustainability indicators 41
Table 4 Sustainability ranking the counties of East Azerbaijan Province 43

Weather Risk Management: The Application of Vine Copula Approach

Table 1 The results of data description 57
Table 2 Results of Kendall's tau, edges, copula families and their estimated parameters in the first tree 58
Table 3 Results of Kendall's tau, edges, copula families and their estimated parameters in the second tree 59
Table 4 Results of Kendall's tau, edges, copula families and their estimated parameters in the third tree 60
Table 5 Results of Kendall's tau, edges, copula families and their estimated parameters in the fourth tree 60
Table 6 Results of copula families and their estimated parameters in D-vine 62
Table 7 Results of vine copula selection 63
Table 8 Theoretical distribution selection for Damavand apple yield. . . . 64
Table 9 Premium in weather-based index insurance for apple 64

An Investigation on Dependency Structure Between Temperature-Humidity Index (THI) and Milk Yield

Table 1 The average monthly THI 82
Table 2 The average monthly milk production per each dairy cow (kg). 83
Table 3 Copula parameters and tail dependency estimation. 84
Table 4 Copula selection 85
Table 5 Theoretical distribution selection 86
Table 6 Expected loss and its probability 86

Adoption of IPM by Farmland Owners and Non-owners: Application of Endogenous Switching Copula Approach

Table 1 Characteristics of farmers in Khuzestan province 99
Table 2 Average willingness to pay of owners in each environmental class 100
Table 3 Average willingness to pay of non-owners in each environmental class 100
Table 4 Results of endogenous switching copula model based on the logit-t-t model 102

Estimating Economic Value and Compensation Surplus of Animal Species in Arasbaran Forests Using Choice Experiment Method

Table 1 The attributes of Arasbaran forests' selected animal species and their levels 114
Table 2 Descriptive statistics of explanatory variables. 117

Table 3 Standard and synthetic estimation conditional logit model for three animal species of Arasbaran forests 117
Table 4 Calculation results of willingness to pay for three selected animal species and their levels 119
Table 5 Results of SC extracting of relative and optimum improvement in three chosen animals' condition 119

Factors Affecting Consumers' Awareness of Pesticides-Free Fruits and Vegetables

Table 1 Variable definition 130
Table 2 Statements of respondent's knowledge of PFFV index. 131
Table 3 Statements of consumer's safer shopping criteria in buying fruit and vegetables 131
Table 4 Statements of respondent's friendly attitude for environment index 132
Table 5 Distribution of AWR variable's responses among households of Marand City. 132
Table 6 Distribution of shop variable's responses among households of Marand City. 133
Table 7 Distribution of ENV variable's responses among households of Marand City. 133
Table 8 Descriptive statistics of respondents' demographic characteristics 134
Table 9 Estimation results of ordered probit model. 134
Table 10 Marginal effects variables from the ordered probit model. 135

The Relationship Between Economic Growth, Energy Consumption, and CO_2 Emissions

Table 1 Summary of related studies 147
Table 2 Unit root test results. 156
Table 3 Co-integration test results. 157
Table 4 Granger causality test results 158

Oil Price Volatility and Food Price Linkage: Evidence of Dutch Disease in Iran's Agricultural Sector

Table 1 Comparison between different estimated volatility models for oil prices. 177
Table 2 Unit root test 179
Table 3 Johansen co-integration test 179

Environmental Efficiency in Agricultural Sector

Table 1 The greenhouse gas emissions in Iran's agricultural sector in 2018 (tons) . . . 184
Table 2 Emission coefficient of carbon dioxide gas . . . 190
Table 3 The data description in Iran's provinces . . . 191
Table 4 The mean silhouette coefficient for different number of clusters . . . 194
Table 5 k-medoids results for clustering provinces based on carbon dioxide emissions caused by agricultural sector . . . 195
Table 6 Results of clusters' agricultural environmental efficiency . . . 196

Introduction to the Economics of Agriculture and Natural Resources

Masoomeh Rashidghalam

1 Background

This volume is a collection of twelve selected empirical studies on the economics of agriculture and natural resources. Twenty-two authors have contributed their research to this volume. Papers of this volume are grouped into three main domains, covering: Agricultural Resilience and Sustainability; Agricultural Producers and Consumers; and Energy Use in Agriculture. Organized in an analytical framework and offering comprehensive empirical data, this book focusses on agricultural sustainability and resilience, environmental efficiency, agricultural extension, foreign trade, energy use, and agricultural growth aspects of the Iranian agriculture sector.

Agriculture is an important and vital sector in Iran's economy. This sector accounts for 20% of the country's GDP and it employs one-third of the workforce. The rapid growth of population and increase in demand for food productions over the past century entailed a dramatic transformation of the traditional farming system. This makes the farmers adopt much more modern production inputs and technologies such as chemical fertilizers and pesticides in the production process. However, overuse and inappropriate use of agrochemicals have led to environmental problems such as contamination of water, loss of genetic biodiversity, and deterioration of soil quality. Considering the importance of these issues in Iran's agriculture, the first part of this volume talks about environmental efficiency, agricultural sustainability, and resilience. The findings have implications for and applicable to problems facing agriculture in Central Asia and MENA region.

The second part of the book consists of different chapters, which investigate consumers' attitudes toward organic and pesticide-free agricultural products. Chemical pesticides are among the most important barriers to sustainability in agriculture.

M. Rashidghalam (✉)
Centre for Entrepreneurship and Spatial Economics (CEnSE), Jönköping International Business School (JIBS), Jönköping, Sweden
e-mail: masoomeh.rashidghalam@ju.se; maso.azar@gmail.com

M. Rashidghalam (ed.), *The Economics of Agriculture and Natural Resources*, Perspectives on Development in the Middle East and North Africa (MENA) Region,
https://doi.org/10.1007/978-981-15-5250-2_1

Their numerous disadvantages for the environment and living organisms, as well as the food cycle, pushed developed societies to eliminate or reduce and imitate pesticide application. Meanwhile, consumers, due to health problems have changed their approach to safer and pesticide-free food products. However, in developing countries such as Iran, lack of knowledge and awareness of healthy food is one of the main reasons for the lack of an organized market for such products. Chapters of this part also discuss various agricultural issues from the producers' perspective. The contributing authors address issues related to insurance and (Integrated Pest Management) IPM adaption. As such, this part of the book offers an invaluable reference guide for academics and practitioners interested in Iran's and similar agriculture structure elsewhere.

The third part of the book deals with energy use in the agricultural sector of Iran. Chapters of this part offer a detailed analysis of the relationship between energy consumption and economic growth. The studies focus on forecasting energy consumption and oil price volatility. These chapters will be of special interest to economists, energy experts, and politicians that deal with energy use issues in agriculture. Finally, by constituting a valuable source of knowledge, this book is important to those managing the agricultural enterprises, policy-makers, and researchers on agricultural producers and consumers.

2 Summary of Individual Studies

Full description of the contributed chapters is presented as the following

2.1 Part One. Agricultural Resilience and Sustainability

The first study (chapter "Quantitative Sustainability Assessment Applied to Dairy Farms") *Quantitative Sustainability Assessment Applied to Dairy Farms* by Leila Hassani, Mahmoud Daneshvar kakhki, Mahmoud Sabuhi sabouni, and Peter Fantke examines the sustainability performance of dairy farms based on water use and energy consumption. In this sense, it uses a multi-indicator modeling approach to evaluate the performance of 30 dairy farms in Khorasan Razavi Province (Iran). The study makes a Sustainability Indicator (SI) using renewable energy, energy intensity, and water usage, which all are considered as the main goal, and SI was obtained using the arithmetic mean of all. This study finds that the value of SI across all farms is about 0.59, which indicates that the sustainability indicator of dairy farms is low in this area. Therefore, it implies that to increase the sustainability level of dairy farms, water and energy consumption must be managed and optimized.

The second study (chapter "A Framework for Economic Resilience Assessment of Agricultural Production Systems") *A Framework for Economic Resilience Assessment of Agricultural Production Systems* by Leila Hassani, Mahmoud Daneshvar kakhki, Mahmoud Sabuhi sabouni, and Peter Fantke tries to evaluate economic resilience and sustainability in industrial dairy cattle farms. Resilience explains how well production systems withstand or rebound from aberration. This includes preserving and restoring agricultural systems under threats that impact national economic development. A dairy farm is a type of agricultural system, production quality and profitability of them are a dominant feature of the economic resilience. These features are used in a multi-indicator modeling approach to assess the economic resilience of 30 dairy farms in Khorasan Razavi Province (Iran). Total revenue and total cost across all farms are 5.18×10^{09} IRR and 2×10^{09} IRR per day, respectively. The average milk quality is 0.52, which indicates low-level milk quality. To evaluate the economic resilience indicator (ERI) for dairy farms, this study develops an economic resilience indicator using profitability and milk quality which are normalized between zero and one. The economic resilience indicator was obtained using the arithmetic mean. The proposed indicator across all farms is about 0.49, meaning that dairy farms are at the low-level. Hence, to increase the economic resilience of dairy farms, it is essential to improve the revenue and production quality of these units while reducing the total cost.

The third study (chapter "Empirical Evaluation of Agricultural Sustainability Using Entropy and FAHP Methods") *Empirical Evaluation of Agricultural Sustainability Using Entropy and FAHP Methods* by Marziyeh Manafi Mollayosefi, Babollah Hayati, Esmaeil Pishbahar, and Javad Nematian using entropy and FAHP techniques attempts to assess the agricultural sustainability in different counties of this province. Comparison of the indicators' weight and rank in FAHP and entropy show that there is a difference between the extracted weights of economic, social, and environmental indicators in two methods. The Efficiency of water consumption and conservation tillage in economic aspect, health, and agricultural employment in social aspect, and efficient irrigation systems and percentage of greenhouse lands in environmental aspect are the most important indicators of both methods. Since the various weighting methods provide different results, the study suggests not to rely solely on the results of an evaluation method, as it may lead to misleading results. The results of this study can aid in the decision-making processes of planning and policy-making. Based on the results, to achieve higher levels of agricultural sustainability, it is recommended to increase the water use efficiency by changing the cropping pattern, accordance with the conditions of the area, and using efficient irrigation systems (e.g., under pressure irrigation systems). The health centers need to have more fair distribution among the counties.

2.2 Part Two. Agricultural Producers and Consumers

The first study (chapter "Weather Risk Management: The Application of Vine Copula Approach") *Weather Risk Management: The Application of Vine Copula Approach* by Sasan Torabi, Arash Dourandish, Mahmoud Daneshvar Kakhki, Ali Kianirad, and Hosein Mohammadi designs the weather-based index insurance for the apple production in Damavand which is considered as an important apple production center in Iran. Data were collected during 1987–2016, from Iranian Agriculture Jihad Organization and the meteorological station in Damavand County. To investigate the dependency structure between weather variables and yield, the C- and D-vine copulas are used, and the Bayesian approach is employed to estimate the copula parameters. Considering the derived expected loss from this dependency structure, the premium is 36,162,340.44 Rial, which is different from the premium of the current insurance. This diversity arises form circular and administrative of current plan and lack of consideration of expected loss in the premium determination.

The second study (chapter "An Investigation on Dependency Structure between Temperature-Humidity Index (THI) and Milk Yield") *An Investigation on Dependency Structure between Temperature-Humidity Index (THI) and Milk Yield* by Afsaneh Nikoukar and Sasan Torabi measures the probability and amount of milk expected loss caused by heat stress in Damavand. Researchers collected the monthly data from Damavand Meteorological and Iranian Agricultural Organization over the period from 2012 to 2016. To increase the accuracy, the dependency structure between milk yield and Temperature-Humidity Index, which is a measurement of heat stress is investigated using the copula function. The evaluation of different families of copula, including Elliptical, Archimedean, and Extreme value, show that there is a strong dependency structure between variables, so that this dependency can be represented by rotated Clayton copula better than the rest. Expected loss within a month for each dairy cow and probability of loss is 42 kg and 8.05%, respectively. Consequently, it is recommended that the Agricultural Insurance Fund implement weather-based index insurance scheme in warm seasons to decrease such losses so that they can stabilize cattlemen's income and protect societies' protein health.

The third study (chapter "Adoption of IPM by Farmland Owners and Non-owners: Application of Endogenous Switching Copula Approach") *Adoption of IPM by Farmland Owners and Non-owners: Application of Switching Copula Approach* by Esmaeil Pishbahar, Sahar Abedi, and Pariya Bagheri investigates the adaption of Integrated Pest Management (IPM) by farmland owners and non-owners. The study employs the endogenous switching copula approach, which allows using different marginal distributions and leads to accurate results. The results show that the logistic distribution for decision equation's residual and Student's t distribution for willingness to pay equation's residual are better than the normal distribution. In addition, the Average Treatment Effect (ATE) results show that owners are willing to pay for IPM a bit more than non-owners. Hence, different factors affect the willingness to pay of the two groups. Owners have more motivation for this kind of operation because they can utilize the long-run benefits. Thus, there is a need for long-run rental contract, and utilization of tax punishment for excessive use of pesticides, which amatively would encourage non-owner to implement IPM.

The fourth study (chapter "Estimating Economic Value and Compensation Surplus of Animal species in Arasbaran Forests Using Choice Experiment Method") *Estimating Economic Value and Compensation Surplus of Animal species in Arasbaran Forests Using Choice Experiment Method* by Maryam Haghjou, Babollah Hayati, Esmaeil Pishbahar, and Morteza Molaei estimates the economic value and compensation surplus (CS) of three valuable animal species of Arasbaran forests: the bear, the tiger, and the black rooster, using choice experiment method and application of conditional Logit regression model. Required data was acquired through field studies and questionnaires filled by 334 visitors and citizens from ten cities in three provinces: East Azerbaijan, West Azerbaijan, and Ardabil. Results showed that total the economic value of these three species is about 692.695×10^9 Rial (About 22×10^6 USD) and 50% of this value is allocated to the tiger. Also, the results revealed that compensating surplus of relative and optimum improvement in animal species' condition would monthly worth about 3,913,333 and 50,733,333 Rial, respectively. Furthermore, the results of this study reveal that respondents' level of education, income, number of annual visits to the forests, and their friendly attitudes toward Arasbaran forests had significant positive impacts on Willingness to Pay of respondents for the animal species.

The fifth study (chapter "Factors Affecting Consumers' Awareness of Pesticides-Free Fruits and Vegetables") *Factors Affecting Consumers' Awareness of Pesticides-free Fruits and Vegetables* by Maryam Haghjou, Babollah Hayati, and Esmaeil Pishbahar examines consumers' awareness of pesticide-free fruits and vegetables and its determinants using Ordered Probit model. Data were collected through the field study and conducted in 2010 among 394 consumers from Marand City (Iran). According to the results, only 20% of the respondents have appropriate information about the features of pesticide-free fruits and vegetables, and about 24% have low information or lack of awareness. Estimation results show that factors such as educational level, positive environmental tendencies, and adherence to healthy lifestyle index among individuals, as well as having children under the age of ten or people with specific diseases in the household have a positive and significant impact on awareness of pesticide-free fruits and vegetables. In this context, female respondents' awareness was more than males. In this respect, appropriate advertising, conducting training courses for all levels of education, raising community awareness of sustainability issues, and safer products besides environmental and a healthy lifestyle issues are suggested.

2.3 Part Three. Energy Use in Agriculture

The first study (chapter "The Relationship Between Economic Growth, Energy Consumption, and CO2 Emissions") *The Relationship between Economic Growth,* Energy Consumption and CO2 Emissions by Parisa Pakrooh and Esmaeil Pishbahar analyzes the relationship between economic growth, energy consumption (including oil, gas, coal, renewable, and electricity), and CO_2 emissions. The important differences of our study are in examining the dynamic relationship among variables concerning volatilities and structural shocks, because of that the TVP-VAR method was applied to this aim during 1978–2015. This method helps to understand the kind of relationship, response of variables to each other's volatilities and structural shocks, and also the size of response, which is necessary for making accurate policies. Also, stationary, co-integration, and causality test are used. Then, the response and the size of responses to structural shocks and fluctuations of variables are analyzed with TVP-VAR model. The results of this study indicate that there is a bidirectional relationship between energy consumption and GDP, and directional relationships between energy consumption and CO_2, also GDP and CO_2. The results of the TVP-VAR model shows that the dynamic Kuznets theory exists, and the size of the response of energy consumption and CO_2 emissions to GDP changes has been lager than other statues. So policy-makers must reduce the energy consumption and CO_2 emissions reactions by replacing renewable energies, which can help to reduce the dependency of sectors to fossil energies and GDP as well as CO_2 reduction.

The second study (chapter "Oil Price Volatility and Food Price Linkage: Evidence of Dutch Disease in Iran's Agricultural Sector") *Oil Price Volatility and Food Price Linkage: Evidence of Dutch disease* by Zahra Rasouli, Mohammad Ghahremanzadeh, and Masoomeh Rashidghalam seeks to establish the relationship between oil price volatility and domestic food price inflation in Iran. Different generalized autoregressive conditional heteroskedasticity (GARCH) type models are estimated to model oil price volatility. Based on the multiple loss functions, periodic GARCH (PGARCH) model is selected as the best. The estimated volatility, together with nominal exchange rate and basic food price inflation are included in a VECM model to estimate the co-integrating vector. The findings reveal that there is a positive and highly significant relationship between food price inflation and oil price volatility and also a negative and significant relation between food price inflation and exchange rate. This long-run relation proves the existence of Dutch disease in Iran's economy.

The third study (chapter "Environmental Efficiency in Agricultural Sector") *Environmental Efficiency in Agricultural Sector* by Sahar Abedi, Pariya Bagheri, and Farid Bagheri Sarajug aims to estimate the agricultural environmental efficiency in Iran's different provinces using data envelopment analysis (DEA) method. The results show that for each billion Rial of agriculture gross value added, 11.39 tons of carbon dioxide are produced. Subsequently, regarding the greenhouse gas emissions caused by proportional agricultural activity, the provinces are divided into three clusters. The mean carbon dioxide emission for each cluster are 536,663.7, 804,315.7, and 186,311.8 tons. The mean environmental efficiency for clusters is 0.294, 0.243, and 0.836, respectively. This indicates that provinces with more greenhouse gas emissions have lower environmental efficiency. In this regard, the politicians should make less efficient clusters to reduce the use of undesirable inputs to decrease pollutant emissions. Besides, by proper allocation of capital, the more efficient cluster can improve environmental technologies.

Masoomeh Rashidghalam is a visiting researcher at Centre for Entrepreneurship and Spatial Economics (CEnSE), Jönköping International Business School (JIBS), Jönköping, Sweden. She did her BSc and PhD in Department of Agricultural Economics at University of Tabriz and holds a MSc from Tarbiat Modares University. Dr. Rashidghalam's areas of expertise are: Agricultural Production Economics, Productivity and efficiency, Well-Being, and Urbanization. She has a wide range of teaching experience in Econometrics, Agricultural Production Economics and Microeconomics. She has written two books: *Measurement and Analysis of Performance of Industrial*

Crop Production: The Case of Iran's Cotton and Sugar Beet Production, 2018, published by Springer. *Sustainable Agriculture and Agribusiness in Iran*, 2019, published by Springer. She has publications in Journal of Productivity Analysis.

Agricultural Resilience and Sustainability

Quantitative Sustainability Assessment Applied to Dairy Farms

Leila Hassani, Mahmoud Daneshvar kakhki, Mahmoud Sabuhi sabouni, and Peter Fantke

Abstract This study examines the sustainability performance of dairy farms based on water use and energy consumption. In this sense, it uses a multi-indicator modeling approach to evaluate the performance of 30 dairy farms in Khorasan Razavi Province (Iran). The study makes a Sustainability Indicator (SI) using renewable energy, energy intensity, and water usage, which are all considered as the main goal, and SI was obtained using the arithmetic mean of all. This study finds that the value of SI across all farms is about 0.59, which indicates that the sustainability indicator of dairy farms is low in this area. Therefore, it implies that to increase the sustainability level of dairy farms, the water and energy consumption must be managed and optimized.

Keywords Energy consumption · Water consumption · Arithmetic mean · Dairy farms · Iran

JEL Classification Q56 · Q25 · Q4

L. Hassani (✉) · M. D. kakhki · M. S. sabouni
Department of Agricultural Economics, Faculty of Agriculture, Ferdowsi University of Mashhad, Mashhad, Iran
e-mail: leila.hassani@mail.um.ac.ir

M. D. kakhki
e-mail: daneshvar@um.ac.ir

M. S. sabouni
e-mail: Sabouhi@um.ac.ir

L. Hassani
Ministry of Agriculture - Jihad, Tehran, Iran

P. Fantke
Division for Quantitative Sustainability Assessment, Department of Management Engineering, Technical University of Denmark, Lyngby, Denmark
e-mail: pefan@dtu.dk

M. Rashidghalam (ed.), *The Economics of Agriculture and Natural Resources*, Perspectives on Development in the Middle East and North Africa (MENA) Region,
https://doi.org/10.1007/978-981-15-5250-2_2

1 Introduction

Water, with its qualitative and quantitative effects, plays an important role in the evolution of human civilization, economic, and social developments (Motiee et al. 2001). Restricted access to water or the presence of contaminants in its supply has a significant impact on the human, plant, animal health, and productivity (Schlink et al. 2010). Water is vital for livestock production because livestock needs water with a similar quality to that required by humans (Schlink et al. 2010). On the other hand, the advancement of technology and the improvement of productivity in dairy farms have been the most critical factors in economic growth and development in Iran over the last few decades (Qobadi and Shahrami 2015). Since 2013, Iran has been facing serious water crisis (Lehane 2014), and in the coming years, energy costs will increase dramatically in agricultural and industrial production processes too (Qobadi and Shahrami 2015). In the field of global competition in production, countries and industries will be more successful in finding ways to prevent waste of water and energy (Qobadi and Shahrami 2015). Although water consumed for milk production, drinking, and preserving dairy cattle is insignificant at global levels (around one percent); it is one of the essential inputs in dry areas especially in Iran (Schlink et al. 2010). Therefore, forecasting water consumption in the livestock industry is of particular importance. The efficiency of water use in livestock production as a component of the human food chain was studied by Schlink et al. 2010. The milk production industry is also one of the most important and energy-intensive agricultural sectors (Qobadi and Shahrami 2015). Farm activities, including feed provision, feeding, milking, and transportation need energy inputs in several forms and various magnitude (Wells 2001). The amount of energy consumption indicators differ between individual farms. Total initial energy input mentions that all forms of energy, measured at the source, for instance, direct energy (fuel, machinery, nutrition, labor, and electricity) and indirect energy (for the production of consumables such as fertilizer) (Wells 2001). Increased consumption of these resources will increase the cost of production (Qobadi and Shahrami 2015) and decrease the sustainability of farms. Due to the significant increase in sustainability knowledge in agricultural production systems, especially in dairy farm (Fiksel 2006), the quantitative sustainability assessment of these units is essential. Sustainability mentions the capability of production systems to continue its action ad infinitum at the future in which the facilities and utility are not depauperate over time. In sustainable dairy farming, reducing greenhouse gas emissions, improving or standardizing water and energy consumption, and decreasing production costs are considered. To determine the sustainability of dairy farms, technical production efficiency and sustainability indicator can be applied. Most of the studies just considered the technical efficiency of dairy farms as sustainability evaluation.

Barnes (2006) assessed the technical efficiency scores of Scottish dairy farms by applying the non-parametric method of Data Envelopment Analysis (DEA) (Barnes 2006). Technical efficiency of Austrian dairy farms was determined using data envelopment analysis (Kirner et al. 2007). Technical efficiency in dairy cattle farms within the province of Izmir (Uzmay et al. 2009) and dairy cattle management in rural areas of the eastern Mediterranean in Turkey (Dagistan et al. 2009) was likewise studied using DEA. In Iran, Qobadi and Shahrami evaluated the energy indices in Qazvin dairy farms using DEA (Qobadi and Shahrami 2015). Sefeedpari et al. (2012) studied energy consumption in dairy farms and assessed the energy efficiency in 2012. The majority of these studies show that the larger farms were more efficient than smaller ones, which is intuitive when exclusively focusing on technical performance. Thompson et al. (1999) and Robinson (2015) concluded that the key to sustainable dairy farms is extensive planning, knowledge of marketing, and the ability to adapt plans as needed (Robinson 2015; Thompson and Nardone 1999). Hassani et al. (2018) evaluated the energy and environmental position of dairy farms in Iran based on standard dairy cattle nutrition (Hassani et al. 2018) and also, in 2019, they introduced a resilience and sustainability indicator for industrial dairy farms via optimization of this indicator and expanding the mathematical programming and metaheuristic algorithms (Hassani et al. 2019).

To best of our knowledge, there are few studies in Iran that focus on the sustainability assessment of industrial dairy farms using sustainability indicators. There is a gap in prior research on the methodology. The majority of them evaluated the technical efficiency using DEA. Therefore, it is the aim of the present study to determine the quantitative sustainability of dairy farms in Khorasan Razavi province of Iran based on energy and water indicators, which are considered as sub-indicators of sustainability. This research tries to design sustainable production processes in the industrial dairy farms, determine the amounts of the produced milk, and also, adjust feed cost, the total cost for each liter of milk production, and energy consumption. The current study could help farmers and decision-makers to improve sustainable manufacturing activities in the dairy farms.

The rest of the paper proceeds as follows. Section 2 deals with the methodology. Section 3 gives the results of the study; the conclusions and policy recommendations are discussed in Sect. 4.

2 Methodology

Khorasan Razavi province with an area of 116485.34 km^2 is the fourth largest province in Iran. This province is located between the geographic coordinates of 33° 52′ and 37° 42′ north latitude from the equator and between 56° 19′ and 61° 16′ east of the Greenwich meridian (Anonymous 2015). For this study, we analyzed the collected data from 30 dairy farms through questionnaires and interviews in 2016 based on non-random sampling. To determine sustainability indicator, dairy farms were evaluated in terms of data for energy and water consumption since dairy farms are energy "consumer" and energy "producer" too (Wells 2001). The input of energy for each farm is calculated by considering their energy content (according to the used sources which are given in Table 1) and related linkage between farm inputs. Then

Table 1 Energy content of inputs in dairy farms

Input energy	Unit	Energy content (MJ/unit)	References
Input			
Labor	H	1.96	Krebs (2002)
Machinery			
a-Weight of tractor	kg	9–10	Gezer et al. (2003)
b-Fix equipment	kg	8–10	Gezer et al. (2003)
c-Electronic motor	kg	64.8	Gezer et al. (2003)
Fuel			
a-Diesel	L	47.8	Gezer et al. (2003)
b-Gas	m^3	49.5	Gezer et al. (2003)
c-Gasoline	L	46.3	Gezer et al. (2003)
Electricity	kWh	11.93	Engineering and Kitani (1999)
Feed			
a-Concentrate	kg	6.3	Komleh et al. (2011)
b-Silage	kg	2.2	Ozkan et al. (2004)
c-Alfalfa	kg	1.5	Meul et al. (2007)
d-Straw	kg	12.5	Qobadi and Shahrami (2015)

they converted all inputs into energy equivalent of inputs to allow for comparing different inputs, including fuels, labor, feed, electricity, and machinery.

We obtained the equivalent energy of each input using the formulas given in Table 2.

The sustainability indicator was developed based on the main objectives of the research, which includes renewable energy, energy intensity, and water intake. Renewable energy (Eq. 1) on dairy farms includes feed and labor energy equivalent (Joung et al. 2013). To estimate the quantity of energy consumed relative to all the other resources input to the system, the energy intensity (Eq. 2) is applied. The production process has less sustainability if more energy is consumed. Hence the ratio is subtracted from 1 and it means minimization of the ratio is guaranteed to gain a high level of sustainability (Joung et al. 2013). Water is among the critical resources used

Table 2 Formulas for calculating equivalent energy inputs of dairy farms

Equivalent energy	Formula	Description
Equivalent energy of fuel consumption	$E_f = F_c \times E_c$	E_f: Equivalent energy of fuel consumption (MJ) F_c: Amount of fuel consumed (L). $E_{c,f}$: Unit fuel energy content (MJ/L)
Equivalent energy of electricity consumption	$E_{\text{elec}} = E_{\text{el}} \times E_{c,\text{el}}$	E_{elec}: Equivalent energy of electricity consumption (MJ) E_{el}: Amount of electricity consumed (kWh) $E_{c,\text{el}}$: Unit electricity energy content (MJ/kWh)
Equivalent energy of labor consumption	$E_{\text{la}} = N_{\text{la}} \times h \times E_c$	E_{la}: Equivalent energy of labor work (MJ) N_{la}: Number of labor workers h : Hours work (h) $E_{c,\text{la}}$: Unit labor energy content (MJ/per person)
Equivalent energy of machinery consumption	$E_m = W_m \times E_c$	E_m: Equivalent energy of machine (MJ). W_m: Mass of machine (kg). $E_{c,m}$: Unit machine energy content (MJ/kg)
Equivalent energy of feed consumption	$E_N = W_N \times E_c$	E_N: Equivalent energy of feed consumption (MJ). W_N: Feed consumed (kg). $E_{c,N}$: Unit feed energy content (MJ/kg)

Sources Qobadi and Shahrami (2015), Sefeedpari (2012)

in production. To calculate this indicator, the daily requirement of dairy cattle and dairy farms based on the standard of the National Council for Agricultural and Natural Resource Research (NRC) was considered to the total water consumption of our sample farms. Required water for dairy cattle is influenced by various factors such as the rate and composition of weight gain, pregnancy, lactation, activity, diet type, feed intake, and ambient temperature (Schlink et al. 2010). The total water intake for livestock is based on different factors such as free water intake, body weight, dry matter intake, dry matter percentage, crude protein, and time. Changes in water intake differ with the physiological state of the livestock body. These changes are significant among species and breeds of dairy cattle. However, when water is calculated based on the body size and the amount of dry matter intake, the differences can be ignored (Schlink et al. 2010). The proposed indicator is calculated as Eq. 3 (Joung et al. 2013):

$$I_{\text{re}} = \frac{\text{Quantity of renewable energy (MJ)}}{\text{Total energy consumption (MJ)}} \tag{1}$$

$$I_{\text{int}} = 1 - \frac{\text{Value of energy consumption (IRR)}}{\text{Value of total inputs to production (IRR)}} \tag{2}$$

$$I_{\text{WU}} = 1 - \frac{\text{Standard water required (here based on NRC)(L/day)}}{\text{Total water consumption (L/day)}} \tag{3}$$

where I_{re}, I_{int}, and I_{WU} are renewable energy, energy intensity, and water usage indicators, respectively, and as sub-indicators of sustainability. All these indicators are between zero and one. If indicators are closer to 1, it indicates that the system is sustainable. Therefore, we get the sustainability indicator using the arithmetic mean.

3 Results and Discussion

Based on the data and information through questionnaires and interviews, we calculated the energy indicators and water usage indicator as shown in Table 3. The results show that total energy consumption for all dairy farms is on average 7.5×10^{08} MJ. According to Table 3, livestock feeds with 4.93×10^{08} MJ contribute to 66% to the total energy consumption, while labor with 6146.56 MJ contribute to only less than 1%, the lowest contribution to overall energy consumption in dairy farms under study. Fuel was the second most energy-consuming input with 25% and 1.9×10^{08} MJ. Fossil fuels such as diesel fuel are used to process animal feed in the farms. Electricity with 6.6×10^{07} MJ and 9% of total energy consumption is the third most energy-consuming input, which is the most used in milking systems, water heaters, and milk coolers. We included for each farm, all machines including tractors, milking, feeding machines which are considered the minimum required equipment. The average renewable energy indicator is 0.45 across dairy farms and shows that the other energies are consumed more than renewable energy. The average energy

Table 3 Result of equivalent energy indicators of dairy farms

Input and output	Equivalent energy (MJ)	Percent (%)	Renewable energy	Energy intensity	Water usage
Input			0.45	0.94	0.39
Fuel	1.9×10^{08}	25.37			
Electricity	6.6×10^{07}	8.8			
Feed	4.9×10^{08}	65.69			
Machinery	1×10^{06}	0.14			
Labor	6146.56	0.0008			
Sum	7.5×10^{08}	100			

Table 4 Determine sustainability indicator Levels of dairy farms in this study

Sustainability level	Sustainability indicator (SI)
Low	SI < 0.6
Middle	0.6 ≤ SI ≤ 0.8
High	SI > 0.8

Source Chaves and Alipaz (2007)

intensity is 0.94, meaning that the ratio of the value energy consumption to the value of total production costs is insignificant. The average water intake was 0.39 across dairy farms, and to improve this ratio, the use of water in dairy farms should be closer to the standards of NRC. The standard form of water requirement based on NRC: on average, the daily water intake of a high-yielding cow is 227.9 L. This amount for medium-yielding cow is 132.9, and for a low yielding cow is 92.9, for calf head is 42.9, for a dry cow (the dairy cattle whose lactation has been finished and waited in turn of Slaughterhouse) is 87.9, and for the dairy farm is 4168.5 L.

After calculating the energy and water consumption indicators and examining the current state of dairy farms, sustainability indicators were obtained. Sustainability indicator through the arithmetic mean of the sub-indicators is calculated as:

$$SI = \frac{I_{re} + I_{int} + I_{WU}}{3} \tag{4}$$

Reducing errors and compensating possible errors is due to the linearity of the structure of Eq. 4 and averaging in separate steps in the calculation process (Chaves and Alipaz 2007). The sustainability indicator is obtained at three levels (scale from zero to one) according to Table 4.

Overall, the sustainability indicator across all sub-indicators is 0.59, which illustrates that the dairy farms have a low level of sustainability.

Previous studies investigated, compared, and determined the efficiency of industrial dairy cattle units, using the data envelopment analysis method. However, in some of these studies, only energy indicators were estimated and none of these studies provided a solution to reduce overall energy consumption in dairy farms. Besides, other studies found that the larger the livestock units are, the higher levels of efficiency can be achieved (Qobadi 2015; Sefeedpari 2012). In line with our results, an overall energy and water consumption can be achieved only by considering per unit of inputs used, how much output is produced, and then combine all aspects in an overall model to arrive at a solution that can effectively be recommended to farmers. Thereby, an input unit is efficient when its production rate is the same as its consumption. To improve the sustainability indicator, in addition to energy and water, it is suggested that other factors affecting the process of improving sustainability are identified, evaluated, and introduced in the overall determination of sustainability indicator. For instance, the policy of producer support estimates (PSE) is one of the most important factors in Iran. Most of the inputs for dairy farms are provided by the government in Iran. In other words, the Iranian Ministry of Agriculture provides livestock feeds through imports and supplies them to dairy farmers at reasonable prices. In fact, it is a kind of producer support estimate. Accordingly, the quality factors of milk production and prices are other factors that can be additionally considered in the assessment of quantitative sustainability in dairy farms. In addition, to consider the energy ratio, net energy gain and productivity energy indicators can be very beneficial in the development of the sustainability of dairy farms.

4 Conclusions and Policy Recommendations

This study was accomplished in Khorasan Razavi province of Iran to assess sustainability indicator based on energy and water consumption in industrial dairy farms. Based on the first part of results, energy consumption is very high, especially related to the cattle feed intake. According to our findings, renewable energy including livestock nutrition needs and increased yields as well as energy consumption, and using water should be minimized in a combined approach of NRC. Appropriate management of using resources and holding encourage plans can be useful in achieving sustainability. Therefore, it is essential that dairy cattle nutrition and water consumption be based on the standard dairy ratio patterns. In other words, it is recommended that the best diet for the dairy cattle be based on the minimum nutrition dairy needed using linear mathematical programming. So, the minimum feed price should first be estimated. In this regard, using modeling to generate alternatives (MGA), the various dietary patterns, which all have the nutritional needs of the livestock are calculated and made available to the livestock farmer so that it can lead to sustainability in the production and optimal use of resources. To achieve a sustainable result, a reduction in water consumption, energy-wasting of feed intake, was considered and consequently increase in production performance, profitability, and sustainability will happen in industrial dairy farms. Given that water needs vary across dairy farms, for example, each dairy cattle has its own water requirement, as well as dairy farms, it is also recommended to extend the software for using water in dairy farms based on the NRC standard. In addition, it is necessary for the government to extend its protectionist policies to the producers of this sector. For instance, to use incentive policies.

References

Anonymous (2015) Annual agricultural statistics. http://www.koaj.ir (in persian)

Barnes A (2006) Does multi-functionality affect technical efficiency? a non-parametric analysis of the Scottish dairy industry. J Environ Manage 80(4):287–294

Chaves HM, Alipaz S (2007) An integrated indicator based on basin hydrology, environment, life, and policy: the watershed sustainability index. Water Resour Manage 21(5):883–895

Dagistan E, Koc B, Gul M, Parlakay O, Akpinar MG (2009) Identifying technical efficiency of dairy cattle management in rural areas through a non-parametric method: a case study for the East Mediterranean in Turkey. J Anim Vet Adv 8(5):863–867

Engineering ICOA, Kitani O (1999) CIGR handbook of agricultural engineering: energy and biomass engineering. American Society of Agricultural Engineers
Fiksel J (2006) Sustainability and resilience: toward a systems approach. Sustain Sci Pract Policy 2(2)
Gezer I, Acaroğlu M, Haciseferoğullari H (2003) Use of energy and labour in apricot agriculture in Turkey. Biomass Bioenerg 24(3):215–219
Hassani L, Daneshvar Kakhki M, Sabouhi M, Ghanbari R, Fantke P (2018). Quantification of energy and environmental aspects of Iranian dairy farms based on optimal nutrition. Int J Adv Sci Eng Techno spl (1):29–34
Hassani L, Daneshvar Kakhki M, Sabouhi M, Ghanbari R (2019) The optimization of resilience and sustainability using mathematical programming models and metaheuristic algorithms. J Cleaner Prod 228(C):1062–10172
Joung CB, Carrell J, Sarkar P, Feng SC (2013) Categorization of indicators for sustainable manufacturing. Ecol Ind 24:148–157
Kirner L, Ortner K, Hambrusch J (2007) Using technical efficiency to classify Austrian dairy farms. Die Bodenkultur 58(1):15–24
Komleh SP, Keyhani A, Rafiee S, Sefeedpary P (2011) Energy use and economic analysis of corn silage production under three cultivated area levels in Tehran province of Iran. Energy 36(5):3335–3341
Krebs J (2002) McCance and Widdowson's the composition of foods: summary edition, 6th edn. The Royal Society of Chemistry/Food Standards Agency, Cambridge/London
Lehane S (2014) The Iranian water crisis. In: Strategic analysis paper. Future Directions International
Meul M, Nevens F, Reheul D, Hofman G (2007) Energy use efficiency of specialised dairy, arable and pig farms in Flanders. Agr Ecosyst Environ 119(1):135–144
Motiee H, Monouchehri G, Tabatabai M (2001) Water crisis in Iran, codification and strategies in urban water. In: Paper presented at the proceedings of the workshops held at the UNESCO symposium, technical documents in hydrology no 45, Marseille, June 2001
Ozkan B, Akcaoz H, Fert C (2004) Energy input–output analysis in Turkish agriculture. Renew Energy 29(1):39–51
Qobadi M, Shahrami A (2015) Evaluation of energy indices in Qazvin dairy farms using data envelopment analysis. Biomed Eng J 4(4):16
Robinson N (2015) Sustainable livestock production. Vet Rec 177(3):i–ii
Schlink A, Nguyen M, Viljoen G (2010) Water requirements for livestock production: a global perspective. Soil Water Manage Crop Nutr Subprogram 6
Sefeedpari P (2012) Assessment and optimization of energy consumption in dairy farm: energy efficiency. Iranica J Energy Environ 3(3):213–224
Thompson P, Nardone A (1999) Sustainable livestock production: methodological and ethical challenges. Livestock Prod Sci 61(2):111–119
Uzmay A, Koyubenbe N, Armagan G (2009) Measurement of efficiency using Data Envelopment Analysis (DEA) and social factors affecting the technical efficiency in dairy cattle farms within the province of Izmir, Turkey. J Anim Vet Adv 8(6):1110–1115
Wells CM (2001) Total energy indicators of agricultural sustainability: dairy farming case. Study final report. Report to MAF policy. Department of Physics, University of Otago

Leila Hassani is an expert of Export Promotion Bureau at Ministry of Agriculture-Jahad, Iran. She did her B.Sc. in Department of Agronomy at Gorgan University of Agricultural and Science and Natural Resources. She holds her M.Sc. from Islamic Azad University—Arsanjan Branch in agricultural economics. She did her Ph.D. in Department of Agricultural Economics at Ferdowsi University of Mashhad, Iran. She was a visiting researcher in Department of Management Engineering at Technical University of Denmark (DTU) during 2017–2018. Dr. Hassani's areas of expertise are agricultural policy and developing, applied econometrics, mathematical programming, sustainability and resilience, marketing and agricultural export. She has a range of teaching experience in agricultural policy, sustainability and general agricultural economic. Her recent publication have appeared in Journal of Cleaner Production. She has written a book: Iranian Trade Exchanges in Agricultural Sector, 2019, published by Agricultural Research, Education and Promotion Organization of Iran. Recently, she has completed the translation of the book "Multilevel Statistical Models", which is under review for publication.

Mahmoud Daneshvar kakhki is a Professor in Department of Agricultural Economics at Ferdowsi University of Mashhad. He holds a B.Sc. from University of Tehran and two M.Sc. in agricultural economics (Sorbonne University, France) and in rural development (University of Montpellier, France). He did his Ph.D. in agricultural economics at Sorbonne University. His areas of interest and research are agricultural policy, rural development, and econometrics. His teaching areas are agricultural policy and development at postgraduate levels.

Mahmoud Sabuhi Sabouni is a Professor of Department of Agricultural Economics at Ferdowsi University of Mashhad, Iran. He holds a B.Sc. in Agricultural Economics from University of Shiraz as well as M.Sc. and his Ph.D. in Agricultural Economics, University of Shiraz. His areas of interest and research are mathematical programming and natural resource economics. His teaching area is microeconomics and mathematical programming at both under- and postgraduate levels. He has over 155 publications in journals and chapters in books.

Peter Fantke is an associate professor in quantitative sustainability assessment with research experience in and focus on assessing fate, exposure, and effects of chemicals and their safer alternatives. He is executive manager of USEtox, the UNEP/SETAC scientific consensus model for characterizing toxicity impacts of chemicals. He contributes to training at M.Sc. and Ph.D. level, organizes international training workshops, initiated and coordinates internal training in his research division, and coordinates global task forces on quantifying emissions of pesticides and addressing health effects from exposure to fine particulate matter and toxic chemicals.

A Framework for Economic Resilience Assessment of Agricultural Production Systems

Leila Hassani, Mahmoud Daneshvar kakhki, Mahmoud Sabuhi sabouni, and Peter Fantke

Abstract This study tries to evaluate economic resilience and sustainability in industrial dairy cattle farms. Resilience explains how well production systems withstand or rebound from aberration. This includes preserving and restoring agricultural systems under threats that impact national economic development. A dairy farm is a type of agricultural system; production quality and profitability of them are a dominant feature of the economic resilience. These features are used in a multi-indicator modeling approach to assess the economic resilience of 30 dairy farms in Khorasan Razavi Province (Iran). Total revenue and total cost across all farms are 5.18×10^{09} IRR and 2×10^{09} IRR per day, respectively. The average milk quality is 0.52, which indicates low-level milk quality. To evaluate the economic resilience indicator (ERI) for dairy farms, we developed an economic resilience indicator using profitability and milk quality which are normalized between zero and one. The economic resilience indicator was obtained using the arithmetic mean. The proposed indicator across all farms is about 0.49, meaning that dairy farms are at the low level. Hence, to increase the economic resilience of dairy farms, it is essential to improve the revenue and production quality of these units while reducing the total cost.

L. Hassani (✉) · M. Daneshvar kakhki · M. Sabuhi sabouni
Department of Agricultural Economics, Faculty of Agriculture, Ferdowsi University of Mashhad, Mashhad, Iran
e-mail: leila.hassani@mail.um.ac.ir

M. Daneshvar kakhki
e-mail: daneshvar@um.ac.ir

M. Sabuhi sabouni
e-mail: Sabouhi@um.ac.ir

L. Hassani
Ministry of Agriculture - Jihad, Tehran, Iran

P. Fantke
Division for Quantitative Sustainability Assessment, Department of Management Engineering, Technical University of Denmark, Lyngby, Denmark
e-mail: pefan@dtu.dk

M. Rashidghalam (ed.), *The Economics of Agriculture and Natural Resources*, Perspectives on Development in the Middle East and North Africa (MENA) Region,
https://doi.org/10.1007/978-981-15-5250-2_3

Keywords Economic resilience indicator · Quality factors in milk · Dairy farms profitability · Arithmetic mean · Dairy farms · Iran

JEL Classification Q01 · L51 · M11 · O10

1 Introduction

Production is a vital requirement for national economic development. In the production process, a set of inputs is converted to outputs. Nevertheless, the transformation process implicates the generation of emissions, inefficient use of available resources, etc. (Galal and Moneim 2015). Therefore, a production system will have a sustained production that is resistant to shocks and hazards (Fiksel 2006). Such a system is resilient when it can anticipate and prevent disasters and recover them over a time (www.fao.org). A resilient economic system is a system that makes effective use of the remaining resources at any given time (Rose 2015). Economic resilience is obtained at three levels of microeconomics, mesoeconomics, and macroeconomics. Microeconomic and Macroeconomic levels are related to individual business and the combination of all economic entities. At the Mesoeconomics level, resilience can reinforce an industry or market, for instance, innovative pricing mechanisms (Rose 2015).

A Dairy farm is a type of agricultural system that has an essential role in societies, providing sources of protein, fertilizer, fuel, etc. Dairy production and impact on natural resources is a dominant feature of livestock for resilience (Naylor 2009). Livestock products not only illustrate a source of high-quality food, but they are an original of revenue for many small farmers in developing countries (www.fao.org). Dairy cattle are fostered to produce milk; hence, milk quality control is an important part of the milk processing industry for all of the scales (small, medium, and large scale). To produce healthy dairy products, raw materials should have good quality. Hence, quality control must begin at the farm. This is achieved through approved methods of milk production. On the other hand, the advancement of technology and the improvement of productivity in dairy farms is one of the most important factors in economic growth and development in Iran over the last few decades. In the coming years, milk production costs will increase dramatically. In the field of global competition in production, countries and industries will be more successful in finding ways to prevent economic loss (Qobadi and Shahrami 2015). van Apeldoorn et al. (2011) studied the resilience of agricultural systems based on Dutch dairy farms and found system models are beneficial. These models recognize destabilizing and stabilizing forces, the changeable variables, and thresholds that assign the resilience of a system (van Apeldoorn et al. 2011). So far, a significant economic resilience indicator is not presented. In most previous studies, the economic resilience indicator is based on guess, assumption, and integration of several variables that are affecting economic resilience. Béné (2013) defined and presented the economic resilience indicator as a resilience cost (Béné 2013). Briguglio et al. (2006) expanded the resilience economic index. They considered macroeconomic stability, microeconomic market efficiency, good governance, and social development as sub-indicators of economic resilience. They evaluated the economic resilience index using the arithmetic mean of those sub-indicators (Briguglio et al. 2006). Darnhofer (2014) assessed the concept of farm management resilience in the face of climate change and economic uncertainty. He illustrated that farm resilience can be augmented or expunged by policy measures and showed how farmers balance the short-term efficiency and the long-term transformability (Darnhofer 2014). Glover (2012) evaluated the resilience of dairy farmers in the UK coping with foot-and-mouth disease (Glover 2012); the resilience of farms in Australia in the face of protracted drought was studied by Sysak (2013) (Sysak 2013). Kenny (2011) examined how to deal with climate change in New Zealand. Darnhofer (2010) studied the strategies of Austrian family farmers to build resilience (Darnhofer 2010). Astigarraga et al. (2011) tested how French beef farmers maintain their resilience in the face of market variations and climatic fluctuations (Astigarraga and Ingrand 2011). In most of these studies, only one feature that would cause resilience was considered. Hassani et al. (2018) measured the energy and environmental aspects of Iranian dairy farms based on optimal nutrition and mathematical programming (Hassani et al. 2018). In another study, they also defined a resilience and sustainability indicator for dairy farms and optimized this indicator by expanding the mathematical programming and metaheuristic algorithms (Hassani et al. 2019).

However, in the study of Hassani et al. (2019), several pillars of sustainability and resilience were integrated. Still, the effect of each pillar alone and its comparison with aggregation was not examined. Thus, based on the existing published literature, studies focusing on the economic resilience assessment of industrial dairy farms according to economic resilience indicators have not been carried out in Iran. Therefore, it is the aim of the present study to evaluate the economic resilience of dairy farms in Khorasan Razavi Province of Iran based on profitability and milk quality indicators, which are some sub-indicators of economic resilience. In sum, the study aims to evaluate the economic resilience of dairy farms based on quality factors and profitability. The results of this study would help the farmers to be more resilient by modifying milk quality and cost.

The rest of the chapter is structured as follows. Section 2 introduces the methodology and the general structure of the model. Section 3 presents the main results of the study, and the final section provides a summary and concluding remarks.

2 Methodology

Khorasan Razavi province is the fourth largest province in Iran and is located in the east north of Iran. (Anonymous 2015). Data used in this study were collected from 30 dairy farms through questionnaires and interviews during 2016 based on non-random sampling. In this non-random sampling method, farms are selected based on the researcher's opinion. In this study, to determine the economic resilience indicator, dairy farms were evaluated in terms of data for quality factors in milk production

Table 1 Economic resilience indicator criteria in dairy farms

Sub-indicator of economic resilience indicator	Criteria description
Quality factors in milk production	a-Somatic Cell Count (SCC) in milk b-Microbial load in milk c-Aflatoxin in milk d-Antibiotics in milk
Profitability	Total cost divided by total revenue

and profitability. The quality factors and profitability of each farm are calculated by considering their criteria, which are given in Table 1.

The Somatic Cell Count (SCC) is the main indicator of milk quality. The majority of them are leukocytes (white blood cells) sent to fight the udder infection. Therefore, the milk from cows that have mastitis has more somatic cells. The Somatic Cell Count is quantified by the number of cells per milliliter of milk. In general terms, the standard of SCC is 100,000 or less and illustrates an "uninfected" cow. The threshold of SCC is 200,000 and would determine whether a cow is infected with mastitis, and the result of greater than 300,000 shows the cow is infected (Schukken et al. 2003).

Bacterial contamination of milk can usually occur after the secretion of milk from the udder by three main sources including outside the udder, within the udder, and from the equipment used for milk handling and storage (Wallac 2009). The microbial load is quantified by colony-forming units per milliliter (CFU/ml) of milk. High bacteria counts (more than 10,000 CFU/ml or more than 100,000 bacteria per milliliter) suggest that bacteria are entering milk from a variety of possible sources (Nádia et al. 2012).

Aflatoxins are the major concern to the dairy industry because they are naturally produced by the fungi and most frequently are found in livestock feed, including corn silage, maize, and cotton seeds. Aflatoxin M1 has a high transfer rate to livestock products such as milk. Fresh milk is continually monitored for aflatoxin M1; if concentrations of M1 is 0.5 mg/kg in the US or more than 0.05 mg/kg in the EU, then they are considered unacceptable and are not used for products that enter the human food chain. In general, based on Food and Drug Administration (FDA), the level of aflatoxin M1 in milk is 0.5 ppb (parts per billion), and such milk is acceptable (Iqbal et al. 2015).

Antibiotics are used to treat cows with mastitis infections, and such cows may have antibiotic residues in their milk. Therefore, these kinds of milk are either discarded or gathered into separate tanks and are not used for human consumption. Therefore, based on the FDA's standard, it is essential that no antibiotics are to be detected on milk while analyzing it (Food and Administration 2009).

More than a decade, the management of product quality has supposed novel strategic importance, giving consumers the right to choose and buy high-quality products at reasonable prices. On the other hand, the quality of production is one of the effective factors in producer support estimate (PSE) by governments (Morgan and Piercy 1998). Therefore, considering these quality factors as sub-indicators of economic resilience indicator seems to be necessary simultaneously with increased profitability in dairy farms.

To calculate microbial load, aflatoxin, antibiotic, and Somatic Cell Count (SCC) indicators, the amount of each factor across all farms considered are based on milk testing and quality control. Then we extended the quality indicator for all of them, which are given in Eqs. 1–4.

The difference between the revenue obtained from product sale and the opportunity cost of the inputs is economic profit (Von Mises 2008). Hence, profitability is the ability of a producer to get profit (Park and Irwin 2004). There are several different ways to measure the profitability of farms. In this study, we focus on profitability ratios of dairy farms, which are measured by dividing total expenditure to total revenue (Eq. 5). Next, the economic resilience indicator is developed based on all aspects of the research, which includes four factors of milk quality and farm profitability.

$$I_{SCC} = \frac{\text{The standard amount of SCC in milk (1000 cell per milliliter)}}{\text{Total amount of SCC in milk across all farms (1000 cell per milliliter)}} \quad (1)$$

$$I_{MC} = \frac{\text{The standard amount of MC in milk (CFU/ml)}}{\text{Total amount of MC in milk across all farms (CFU/ml)}} \quad (2)$$

$$I_{AF} = \frac{\text{The standard amount of Aflatoxin in milk (ppb)}}{\text{Total amount of Aflatoxin in milk across all farms (ppb)}} \quad (3)$$

$$I_{AN} = \frac{\text{The standard amount of Antibiotic in milk (mg/kg)}}{\text{Total amount of Antibiotic in milk across all farms (mg/kg)}} \quad (4)$$

$$I_{Profit} = \frac{\text{Total Cost (IRR)}}{\text{Total Revenue (IRR)}} \quad (5)$$

where I_{SCC}, I_{MC}, I_{AF}, I_{AN}, and I_{Profit} are Somatic Cell Count indicator, Microbial load indicator, Aflatoxin indicator, Antibiotic indicator, and profitability indicators, respectively. In this research, labor and farm manager salary, feed cost, fuel cost, sperm, and inoculation are considered as the total cost. Total revenue is obtained by the sale of products such as milk, calf, manure, and meat. All five indicators are sub-indicators of economic resilience and will be between zero and one. As the result closes to 1, it shows that the system is resilient. Finally, the economic resilience indicator is obtained using the arithmetic mean.

3 Results and Discussion

The current quality and profitability of the dairy farms were calculated via data extracted from the questionnaires and interviews which is shown in Table 2. The results show that the total Somatic Cell Count in milk for all dairy farms is on average 269,500 Cells per ML. This means that more than half of the cows in our study are infected by mastitis. On the other hand, SCC monitoring is important for dairy producers because a lower SCC indicates high-grade animal health. If the number of somatic cells rises, milk yield is likely to fall or has poor quality or even leads to unsaleable milk. Such milk is inappropriate for human consumption. The same as SCC, total microbial load and aflatoxin in milk across all farms are 283,000 bacterial per ml and 224 ppb, respectively, which highlights low milk quality in the region. In addition to SCC, the microbial load is also obtained from cows with mastitis. Increased microbial load in raw milk has consequences including fat, protein, and milk glucose corruption, the reduced production efficiency of dairy products such as cheese and yogurt, etc. As well as mycotoxin in livestock feed led to animal productivity reduction and contamination of cattle products such as milk, meat, etc., Aflatoxins is a kind of toxins that causes the economic losses in this area. Aflatoxin B1 is the most strong of the aflatoxin group. It has a carcinogen factor, and aflatoxin M1 is a product of aflatoxin B1 and appears in the milk of lactating cows (Iqbal et al. 2015). Based on this the FDA's regulate the level of aflatoxin in milk is 0.5 ppb; however, in our study the level is obtained as 224 ppb and indicates the detection is necessary. Also, the control of aflatoxin should be expanded seriously in this field because it costs human and animal health. According to Table 2, 15 mg/ml antibiotic is detected across all farms, and it seems logical because antibiotics are used to treat cows with mastitis infections. According to our results, highly, likely half of the cows in dairy farms of this region have mastitis.

Table 2 Result of quality factors and profitability indicators of Dairy Farms

Indicator	Amount	Unit	SCC	Microbial load	Aflatoxin	Antibiotic	Profitability
Quality			0.42	0.48	0.17	0.98	0.46
SCC	269,500	Cell/ml					
Microbial load	283,000	CFU/ml					
Aflatoxin	224	Ppb					
Antibiotic	15	Mg/ml					
Profitability							
Total cost	2×10^{09}	IRR/day					
Total revenue	5.2×10^{09}	IRR/day					

Due to the low level of production quality, 0.52, as expected these dairy farms have the lowest profitability. According to the results, the profitability ratio across all farms is 0.46. The profitability of a production unit is related to production costs. In dairy farms, 60–70% of the costs are related to animal feed. Therefore, it is essential to manage economical and high-quality livestock feeding in dairy farms. It should be noted that quality factors of milk, as well as profitability of each dairy farm are calculated individually. The results indicated that 6 dairy farms (20%), 2 dairy farms (6.7%), 4 dairy farms (13.3%), 2 dairy farms (6.7%), and 15 dairy farms (50%) from 30 dairy farms under study have a middle-level of profitability, Somatic Cell Count in milk, Microbial load in milk, Aflatoxin in milk, and Antibiotics in milk, respectively. However, it does not imply that these farms are at the level of economic resilience because by assessing the average of quality and profitability indicators or economic resilience indicator for each farm, it was observed that none of them are at the resilient level.

Then, after calculating the quality factors, profitability indicators, and examining the current state of dairy farms, economic resilience indicators are obtained. Economic resilience indicator (ERI) through the arithmetic mean of the sub-indicators is calculated as

$$\mathrm{ERI} = \frac{I_{\mathrm{SCC}} + I_{\mathrm{MC}} + I_{\mathrm{AF}} + I_{\mathrm{AN}} + I_{\mathrm{Profit}}}{5} \tag{6}$$

Compensating errors reduced due to the linearity of Eq. 6 and average in separate steps in the calculation process. (Chaves and Alipaz 2007). The economic resilience indicator is obtained at three levels (scale from zero to one) according to Table 3.

Overall, the economic resilience indicator across all sub-indicators is 0.49, indicating that dairy farms in this study have low-level economic resilience. This result is in line with the results of the studies by Hassani et al. (2018, 2019). Previous studies investigated, compared. and determined the economic resilience of industrial dairy cattle units, using the cost, microeconomic market efficiency, policy measurement, and cattle diseases. However, in some of these studies, only the cost of milk production was estimated, but most of these studies did not provide any solution to reduce the overall cost in dairy farms. Hassani et al. (2019) assessed the resilience and sustainability indicator in the field of all dimensions. They concluded that if all the components of sustainability and resilience are considered together, it is possible to

Table 3 Determined economic resilience indicator levels of dairy farms in this study

Economic resilience level	Economic resilience indicator (ERI)
Low	ERI < 0.6
Middle	$0.6 \leq$ ERI ≤ 0.8
High	ERI > 0.8

Source Chaves and Alipaz (2007)

achieve a high-performance production system. In line with our results, by considering per unit of inputs used, total cost, revenue, and quality factors of milk production can be achieved. For instance, how much output is produced? Then combine all aspects in the final model to arrive at a solution that can effectively be recommended to farmers. Thereby, an input unit is efficient when its production rate is the same as its consumption. To improve the economic resilience indicator, in addition to profitability and milk quality, it is suggested that other factors affecting the process of improving economic resilience are to be identified, evaluated, and introduced in the overall determination of economic resilience indicator. For instance, the policy of producer support estimates (PSE) is one of the most important factors in this regard in Iran. Milk demand function, the risk or uncertainty and set prices accordingly are other factors that can be additionally considered in the assessment of economic resilience and sustainability in dairy farms. Reviewing each goal should be based on its priority, and also can be effective in changing the units. In this regard, it is recommended to use a weighted geometric mean. In the present study, as well as previous studies, all data have been conducted on the basis of cross-sectional data. Therefore, in addition to the above mentioned, it is recommended to use the dynamic method.

4 Conclusions and Policy Recommendations

This study was accomplished in Khorasan Razavi Province of Iran to assess the economic resilience indicator based on milk quality and profitability in industrial dairy farms. Based on the results, quality factors including Somatic Cell Count, Microbial load, Aflatoxin, and Antibiotic in milk are very high, especially Aflatoxin. The limited attention and supervision to the quality of animal feed, animal health, and animal sanitation are the main reasons for economic losses in dairy farms, which in the long term leads to the elimination of farms. According to our findings, economic restraint in dairy farms will be achieved through three points: first, to concentrate on animal health and sanitation; Second, the quality of milk production; and based on these aspects, the resilience of economics in dairy farms will reach the global standard. It means the amount of Somatic Cell Count, Microbial load, Aflatoxin, and Antibiotics in milk should be under the FDA regulation. Finally, implementing encouraging policies to support the producers. For instance, setting the price of dairy products (milk, yogurt, cheese, etc.) should be based on their quality. In this approach, the producers are encouraged to produce a higher quality product to achieve greater profitability, and the consumer through having the right to choose goods can guarantee his/her health. Therefore, appropriate management of holding encourages plans can be effective in achieving economic resilience. To achieve the optimum result of a resilient economy, increasing animal health must be considered, and thereafter, we will see increased production performance, profitability, and economic resilience in industrial dairy farms.

References

Anonymous (2015) Annual agricultural statistics. http://www.koaj.ir (in persian)
Astigarraga L, Ingrand S (2011) Production flexibility in extensive beef farming systems. Ecol Soc 16(1)
Béné C (2013) Towards a quantifiable measure of resilience. In: IDS working papers, vol 434, pp 1–27
Briguglio L, Cordina G, Farrugia N, Vella S (2006) Conceptualizing and measuring economic resilience. In: Building the economic resilience of small states, Malta: Islands and Small States Institute of the University of Malta and London: Commonwealth Secretariat, pp 265–288
Chaves HM, Alipaz S (2007) An integrated indicator based on basin hydrology, environment, life, and policy: the watershed sustainability index. Water Resour Manage 21(5):883–895
Darnhofer I (2010) Strategies of family farms to strengthen their resilience. Environ Policy Gov 20(4):212–222
Darnhofer I (2014) Resilience and why it matters for farm management. Eur Review of Agri Econ 41(3):461–484
Fiksel J (2006) Sustainability and resilience: toward a systems approach. Sustain Sci Pract Policy 2(2)
Food U, Administration D (2009) Grade “A” pasteurized milk ordinance. US Food and Drug Administration, Washington, DC, p 52
Galal NM, Moneim AFA (2015) A mathematical programming approach to the optimal sustainable product mix for the process industry. Sustainability 7(10):13085–13103
Glover J (2012) Rural resilience through continued learning and innovation. Local Econ 27(4):355–372
Hassani L, Daneshvar Kakhki M, Sabouhi M, Ghanbari R, Fantke P (2018) Quantification of energy and environmental aspects of Iranian dairy farms based on optimal nutrition. Int J Adv Sci Eng Techno spl (1):29–34
Hassani L, Daneshvar Kakhki M, Sabouhi M, Ghanbari R (2019) The optimization of resilience and sustainability using mathematical programming models and metaheuristic algorithms. J Cleaner Prod 228(C):1062–10172
Iqbal SZ, Jinap S, Pirouz A, Faizal AA (2015) Aflatoxin M 1 in milk and dairy products, occurrence and recent challenges: a review. Trends Food Sci Technol 46(1):110–119
Kenny G (2011) Adaptation in agriculture: lessons for resilience from eastern regions of New Zealand. Clim Change 106(3):441–462
Morgan NA, Piercy NF (1998) Interactions between marketing and quality at the SBU level: influences and outcomes. J Acad Mark Sci 26(3):190–208
Nádia M, Diane S, Débora O, Mirlei RE (2012) Evaluation of microbiological quality of raw milk produced at two properties in the far west of Santa Catarina, Brasil. Food Public Health 2(3):79–84
Naylor RL (2009) Managing food production systems for resilience. In: Principles of ecosystem steward. Springer, pp 259–280
Park CH, Irwin SH (2004) The profitability of technical analysis: a review
Qobadi M, Shahrami A (2015) Evaluation of energy indices in Qazvin dairy farms using data envelopment analysis. Biomed Eng J 4(4):16
Rose A (2015) Measuring economic resilience: recent advances and future priorities
Schukken YH, Wilson DJ, Welcome F, Garrison-Tikofsky L, Gonzalez RN (2003) Monitoring udder health and milk quality using somatic cell counts. Vet Res 34(5):579–596
Sysak TS (2013) Drought, power and change: using Bourdieu to explore resilience and networks in two northern Victoria farming communities. University of Melbourne, Department of Resource Management and Geography
van Apeldoorn D, Kok K, Sonneveld M, Veldkamp T (2011) Panarchy rules: rethinking resilience of agroecosystems, evidence from Dutch dairy-farming. Ecol Soc 16(1)
Von Mises L (2008) Profit and loss. Ludwig von Mises Institute
Wallace RL (2009) Bacteria counts in raw milk. Dairy Cattle Manage 1–4
www.fao.org. (2015)

Leila Hassani is an expert of Export Promotion Bureau at Ministry of Agriculture-Jahad, Iran. She did her B.Sc. in Department of Agronomy at Gorgan University of Agricultural and Science and Natural Resources. She holds her M.Sc. from Islamic Azad University—Arsanjan Branch in agricultural economics. She did her Ph.D. in Department of Agricultural Economics at Ferdowsi University of Mashhad, Iran. She was a visiting researcher in Department of Management Engineering at Technical University of Denmark (DTU) during 2017–2018. Dr. Hassani's areas of expertise are agricultural policy and developing, applied econometrics, mathematical programming, sustainability and resilience, marketing and agricultural export. She has a range of teaching experience in agricultural policy, sustainability and general agricultural economic. Her recent publication have appeared in Journal of Cleaner Production. She has written a book: Iranian Trade Exchanges in Agricultural Sector, 2019, published by Agricultural Research, Education and Promotion Organization of Iran. Recently, she has completed the translation of the book "Multilevel Statistical Models", which is under review for publication.

Mahmoud Daneshvar kakhki is a Professor in Department of Agricultural Economics at Ferdowsi University of Mashhad. He holds a B.Sc. from University of Tehran and two M.Sc. in agricultural economics (Sorbonne University, France) and in rural development (University of Montpellier, France). He did his Ph.D. in agricultural economics at Sorbonne University. His areas of interest and research are agricultural policy, rural development, and econometrics. His teaching areas are agricultural policy and development at postgraduate levels.

Mahmoud Sabuhi Sabouni is a Professor of Department of Agricultural Economics at Ferdowsi University of Mashhad, Iran. He holds a B.Sc. in Agricultural Economics from University of Shiraz as well as M.Sc. and his Ph.D. in Agricultural Economics, University of Shiraz. His areas of interest and research are mathematical programming and natural resource economics. His teaching area is microeconomics and mathematical programming at both under- and postgraduate levels. He has over 155 publications in journals and chapters in books.

Peter Fantke is an associate professor in quantitative sustainability assessment with research experience in and focus on assessing fate, exposure, and effects of chemicals and their safer alternatives. He is executive manager of USEtox, the UNEP/SETAC scientific consensus model for characterizing toxicity impacts of chemicals. He contributes to training at M.Sc. and Ph.D. level, organizes international training workshops, initiated and coordinates internal training in his research division, and coordinates global task forces on quantifying emissions of pesticides and addressing health effects from exposure to fine particulate matter and toxic chemicals.

Empirical Evaluation of Agricultural Sustainability Using Entropy and FAHP Methods

Marziyeh Manafi Mollayosefi, Babollah Hayati, Esmaeil Pishbahar, and Javad Nematian

Abstract Over the last decades, sustainable agriculture is presented as the primary alternative to conventional farming in the world, reacting against a set of problems related to environmental natural resource issues. Evaluating agricultural sustainability is a crucial component for improving agricultural sustainability. East Azerbaijan Province is one of the agricultural centers of Iran. This study, using entropy and FAHP techniques, attempts to assess the agricultural sustainability in different counties of this province. Comparison of the indicators' weight and rank in FAHP and entropy show that there is a difference between the extracted weights of economic, social, and environmental indicators in two methods. Efficiency of water consumption and conservation tillage in economic aspect, health and agricultural employment in social aspect, and efficient irrigation systems and percentage of greenhouse lands in environmental aspect are the most important indicators of both the methods. Since the various weighting methods provide different results, we suggest not to rely solely on the results of an evaluation method, as it may lead to misleading results. The results of this study can aid in the decision-making processes of planning and policy-making. Based on the results, to achieve higher levels of agricultural sustainability, it is recommended to increase the water use efficiency by changing the cropping pattern, accordance with the conditions of the area, and using efficient irrigation systems (e.g., under pressure irrigation systems). The health centers need to have more fair distribution among the counties.

M. Manafi Mollayosefi (✉) · B. Hayati · E. Pishbahar
Department of Agricultural Economics, University of Tabriz, Tabriz, Iran
e-mail: m.manafi.m@tabrizu.ac.ir

B. Hayati
e-mail: b-hayati@tabrizu.ac.ir

E. Pishbahar
e-mail: pishbahar@tabrizu.ac.ir

J. Nematian
Department of Industrial Engineering, University of Tabriz, Tabriz, Iran
e-mail: Jnematian@tabrizu.ac.ir

M. Rashidghalam (ed.), *The Economics of Agriculture and Natural Resources*, Perspectives on Development in the Middle East and North Africa (MENA) Region,
https://doi.org/10.1007/978-981-15-5250-2_4

Keywords Entropy · Evaluation · Fuzzy analytic hierarchy process (FAHP) · Indicators of sustainability · Sustainable agriculture

1 Introduction

The rapid population growth and increasing demand for food over the past century entailed a dramatic transformation of the traditional farming system. This made the farmers use much more modern inputs such as chemical fertilizers and pesticides in production activities, but overuse and inappropriate use of agrochemicals have led to environmental problems such as water pollution, low genetic diversity, and soil degradation (Iravani and Darban-Astaneh 2004). For the same reason, in the recent decades, new attitudes toward sustainable and sound utilization of natural resources are formed (Pretty 1995). Sustainable development principles were introduced by the Brundtland Report (UNWCED UNWCED 1987) for the first time. This report points out the importance of sustainable and equitable natural resources' utilization within and among generations at the global level. The agricultural sector is critical for sustainable development goals worldwide. Sustainable agriculture is a new approach, and constitutes an alternative to traditional and industrial agriculture. While many definitions and several explanations exist, there is general agreement about the description of "sustainable agriculture" as an activity that constantly satisfies certain conditions for an unlimited time (Hansen 1996).

Sustainability performance evaluation is one of the most fundamental issues in agricultural economy, because it is a crucial component for improving agricultural sustainability. For any study on sustainable agriculture, the question arises as to how to evaluate the agricultural sustainability. There are extensive methods and indicators for evaluating the agricultural sustainability, but those that integrate and aggregate the collected information to manage and to document sustainability improvements are of great interest, not only to farmers but also to policy-makers and other stakeholders. Thus, developing a composite indicator (CI) that collects information from these extensive sustainability assessments seems particularly useful (Dong et al. 2015).

Many studies have focused on the assessment of agricultural sustainability all over the world, and have made considerable efforts to identify appropriate indicators and weighting methods evaluation of the agricultural sustainability. Qian and Xueping (2007) assessed the agricultural sustainability of Shaanxi Province of China from 1994 to 2003. They used 33 indicators in five dimensions of population, economy, society, resource, and environment. Researchers used Shannon Entropy method to weight the indicators. According to the results, the indicators such as educational level, the per capita under cultivation, and consumption of pesticides and chemical fertilizers per hectare were the main factors to restrict agricultural sustainability. Gomez-Limon and Sanchez-Fernandez (2010) developed a practical methodology to evaluate farms' sustainability using composite indicators. They applied it to two agricultural systems in Spain. Furthermore, they used 16 sustainability indicators that cover the three components of the sustainability (economic, social, and environmental), and two weighting methods (AHP and PCA) to make the composite indicators. The results showed that the application of such techniques could improve the sustainability of the sector. Ceyhan (2010) assessed the sustainability of agricultural activities in Samsun Province of Turkey using the AHP method. To identify the sustainability level, he used total sustainability index, based on 40 sustainability indicators. The results revealed that there were some severe problems in the social, economic, environmental, and biophysical aspects of agricultural sustainability as land ownership, inadequate health service, insufficient sewerage systems, low level of return on an asset, excessive chemical input use, insufficient irrigation water, and water erosion. In another study, Radulescu et al. (2010) employed a hybrid of AHP and TOPSIS (Technique for Order of Preference by Similarity to Ideal Solution) method to evaluate the performance of agricultural sustainability in Romania. According to the results, the economic aspect among the criteria and the average yield among the sub-criteria had the most effect on agricultural sustainability.

The agricultural sustainability of selected counties of the Fars Province in Iran was discussed by Pourzand and Bakhshode (2012). They used the compromise programming approach in their evaluation. The results showed that higher levels of nitrate concentration in groundwater, negative groundwater balance, limited utilizing of efficient irrigation systems, and chemical fertilizers and pesticides are specific features of the unstable group. Koocheki et al. (2013) used the relative advantage method to assess the agricultural sustainability in the provinces of Iran. They considered five indices including agricultural resources, agriculture development, environment, rural societies, and education based on 60 indicators. The results showed that Iran had unsustainable or weak sustainable status in the development of sustainable agriculture because only five provinces from 30 provinces had high and moderate sustainability, and others had low sustainability or un-sustainability. Dong et al. (2015) used a composite indicator to measure the farm sustainability in Wisconsin Cranberry of the USA. The combination of non-negative polychoric principal component analysis and data envelopment analysis (DEA) has been used to individually score each farm. The results showed heterogeneity in sustainability practice adoption, implying that some farms could adopt relevant practices to improve the overall sustainability performance of the industry. Jamali Moghaddam et al. (2017) evaluated the agricultural sustainability of Pasargad plains' farmers in Fars Province of Iran. They used principal component analysis and hierarchical analysis methods for weighting the indicators and compared the results of them. The results showed that there was a statistically significant difference between economic and social sustainability indices in two weighting methods. However, among environmental sustainability indicators, there was no statistically significant difference based on AHP and PCA. Overall, there was no statistically significant difference between the two methods based on a combination of sustainability indicators. Literature review showed that many studies have been conducted to evaluate agricultural sustainability using different methods and at different levels of measurement such as farm, county, and province. Based on the review of previous studies, the sustainability assessment method and the used indicators in this study were selected.

In Iran, the issue of sustainable agriculture and its development has been considered by long-run programming such as fourth and fifth 5-year development programs, and there are many policies in this area. Nevertheless, Iran is weak in terms of sustainability in agriculture resources, environment, rural society, and agriculture education (Koocheki et al. 2013). East Azerbaijan Province is one of the agricultural centers of Iran. The latest Iranian national census of agriculture in 2014 indicated that this province was ranked three and five in terms of agricultural land area and the number of farmers among the other provinces, respectively. East Azerbaijan Province produces six percent of Iran's total agricultural production by consuming three percent of agricultural water, three percent of chemical fertilizers, and two percent of total chemical pesticides of the country (Statistical Center of Iran 2014). East Azerbaijan Province is an interesting place for evaluation of sustainability because of many problems in achieving the agricultural sustainability. Low yield, intense soil erosion, inefficient irrigation systems, poor financial support, and pasture overexploitation are examples of these problems. Because of these basic obstacles, the agricultural sustainable development in this province is weak. (Jihad-agriculture organization of East Azerbaijan province 2015). Therefore, the evaluation of agricultural sustainability can illustrate the sustainability status in different counties of this province, and also provide the comparison and identification of advantage and disadvantage properties of them and help us to find the problems in achieving the agricultural sustainability and solve them. On the other hand, literature review indicated that few studies have been done on assessing the agricultural sustainability in this province. Therefore, it is necessary to conduct studies in this province to assess the agricultural sustainability and identify its determinants. So this study aims to evaluate the agricultural sustainability of East Azerbaijan Province (Iran). The findings of this research can be used to improve the current policies of the province in the agriculture sector (such as water use policy, rural development policy, and agricultural structure policy) and upgrade them.

The rest of this paper is organized as follows. Section 2 deals with the methodology which is followed by findings of the study. The last section gives the conclusion and policy recommendation.

2 Methodology

Due to its complex and multi-dimensional nature, agricultural sustainability is most often assessed using numerous indicators. The main feature of indicators is their ability to summarize, focus, and condense the enormous complexity of dynamic environment to a manageable amount of meaningful information (Kumar Singh et al. 2010). Evaluating of agricultural sustainability using indicators has considerable shortcomings. A particular difficulty lies in the interpretation of the whole set of indicators. This makes the concept difficult to communicate to the public, policy-makers, and the media. While extensive assessments and indicators exist that reflect the different facets of agricultural sustainability, because of the relatively large number of measures and interactions among them, a composite indicator that integrates and aggregates the overall variables is particularly useful (Dong et al. 2015). The composite indicator is recognized as a helpful tool for evaluating complex and sometimes vague and elusive concepts such as sustainability, environmental performance, and policy analysis. Generally, a CI is the mathematical combination of individual indicators based on an underlying model, taking into consideration methodological assumptions and subjective as well as objective judgments (Roy et al. 2014).

Literature review shows that there are many ways to make a composite indicator. In this paper, we use the weighted sum of normalized indicators, as a representative of linear additive methods (total compensation among indicators). The summation of weighted and normalized indicators is the most popular method for making the composite indicators and is widely used in the studies of sustainability evaluation. From the mathematical point of view, this is a linear weighted aggregation rule applied to the set of normalized indicators (OECD-JRC 2008):

$$\text{CIAS} = \sum_{k=1}^{k=n} W_k.I_k \tag{1}$$

where CIAS is the composite indicator of agricultural sustainability, W_k is the weight associated with the indicator k, and I_k is the normalized value of indicator k. In this case, among the various normalization techniques available (Freudenberg 2003), we decided to employ "min–max" normalization. In this technique, the values of all the normalized indicators vary within a dimensionless range of 0–1, where 0 corresponds to the worst possible value of the indicator (i.e., the least sustainable) and 1 to the best (most sustainable).

The weighting techniques for constructing composite indicators are divided into "positives" and "normative" (OECD-JRC 2008). The positive or endogenous techniques allow us to obtain the weights of the base indicators via statistical procedures, without having to include value judgments of their relative importance. Normative or exogenous techniques attempt to assign different weights to the indicators as a function of the opinion of experts and external decision-makers. Using this type of weighting is similar to introducing social preferences regarding individual dimensions of sustainability (sustainability as a social construction) into the analysis (Gomez-Limon and Sanchez-Fernandez 2010).

The selection of a particular technique of weighting indicators may influence the result of the composite indicator. For this reason, we use two weighting methods to evaluating the sustainability. Hence, we use Principal Components Analysis (PCA), as a method representative of the positive approach, and Fuzzy Analytic Hierarchy Process (FAHP) as a representative of the normative approach.

Entropy was introduced into information theory by Shannon in 1948. The entropy method is an objective weighting method based on the principle that greater uncertainty about outcomes results in a more uniform probability assigned to them. Nowadays, this method has been widely used in research on evaluation studies (Ding et al. 2016). Theoretically, the calculation steps are (Shannon 1948)

To calculate the entropy value of each indicator, the standardized value of indicator j for object i must first be calculated, and is written as Eq. (2):

$$P_{ij} = \frac{X_{ij}}{\sum_{i=1}^{n} X_{ij}} \tag{2}$$

where X_{ij} is the value of the j th indicator of the ith object (in this study countries). Then, calculate the entropy value of the jth indicator using Eq. (3):

$$E_j = -M \sum_{i=1}^{n} P_{ij} \ln P_{ij} \tag{3}$$

M is a constant equal to $(\ln n)^{-1}$, which assures $0 \leq E_j \leq 1$ and $P_{ij} \ln P_{ij} = 0$ if $P_{ij} = 0$. The larger the E_j, the less information is transmitted by the j th criterion. The weighting value of the j th indicator is defined as

$$W_j = \frac{1 - E_j}{\sum_{j=1}^{m} 1 - E_j}; \forall j \tag{4}$$

The Analytic Hierarchy Process (AHP) for choosing factors that are important for decision-making (DM) was proposed by Saaty (1980). This is one of the useful methods in multi-criteria decision-making (MCDM), which has found wide application in many areas of science and practice. The FAHP technique can be viewed as an advanced analytical method developed from the traditional AHP. A Fuzzy set theory provides a strict mathematical framework in which vague conceptual phenomena can be precisely and rigorously studied. This technique is a suitable tool to reinforcement the comprehensiveness and correctness of the decision-making stages (Aggarwal and Singh 2013). In this paper, we use Chang's extent analysis method for the evaluation of agricultural sustainability in the counties of East Azerbaijan Province (Iran). The principles of the FAHP in this method are as follows (Chang 1992):

(1) In the first step, we define the problem and create a hierarchical structure (Saaty 1980). Figure 1 shows one model with three levels for evaluating the sustainability. Level 0 is related to the overall goal, which includes ranking of alternatives and determination of the best or most appropriate alternative. Level 1 encompasses prescribed criteria and sub-criteria, and level 2 contains alternatives that are related to these criteria.

(2) In the second step, we prepare a pairwise comparison questionnaire to collect the experts' opinions on the importance of the criteria and sub-criteria. In FAHP, common sense linguistic statements have been used in the pairwise comparison

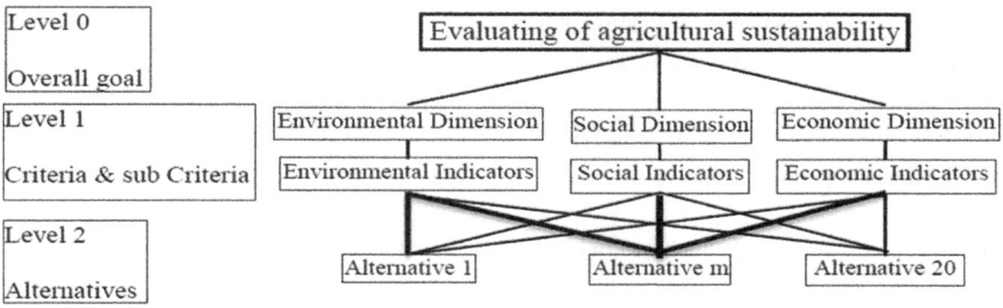

Fig. 1 Hierarchical levels

which can be represented by the triangular fuzzy numbers (TFNs). According to the responses on the question forms, the corresponding triangular fuzzy values are placed for the linguistic variables, and the pairwise comparison matrix is constructed for a particular level on the hierarchy.

(3) In this step, we estimate the weights of indicators based on Chang's extent analysis method (1992).

(4) Since the comparisons are carried out through personal or subjective judgments, some degree of inconsistency may occur. To guarantee that the judgments are consistent, the final operation called consistency verification is incorporated in order to measure the degree of consistency among the pairwise comparisons by computing the consistency ratio. If consistency ratio exceeds the limit, the decision-makers should review and revise the pairwise comparisons (Aggarwal and Singh 2013).

The selection of proper indicators is a necessary principle of this kind of study. In this respect, using an extensive review of the literature, we select a set of agricultural sustainability indicators in each dimension. Then, consulting with a group of 28 experts and researchers (seven university researchers, seven academic members of agricultural research centers, seven Agriculture-Jihad experts, three experts of environment organization, and two experts of natural resources organization), we choose useful and measurable indicators. The members of this group were chosen due to their skill, competence, and knowledge in the sustainable agricultural activities, then were asked to select the best sustainability indicators based on the important criteria of indicators selection such as scientifically precise, measurability, sensitivity to changes, and economically affordable (Sauvenier et al. 2006). According to the experts' viewpoints, 29 indicators were selected in three dimensions. Table 1 presents the list of indicators for the evaluation of agricultural sustainability. The other required data and information were collected from Agriculture Jihad organization's statistical reports, statistical yearbooks of East Azerbaijan Province, and analytical reports of Regional Water Organization of East Azerbaijan Province in 2015.

Table 1 List of indicators for evaluation of agricultural sustainability

Economic sustainability	Social sustainability	Environmental sustainability	
Insurance percentage	Health indicator	Groundwater level	Water salinity
Irrigated wheat yield	Population density	Percentage of greenhouse Lands	Soil organic matter
Efficiency of water consumption	Connection indicator	Efficient irrigation system	Fallow indicator
Economic participation rate	Literacy rate	Agricultural water use	Vegetation cover
Mechanization level	Extension activities	Indicator of pesticide usage	Grazing indicator
Agricultural land per capita	Agricultural cooperatives	Indicator of fertilizer use	nitrate Concentration
Conservation tillage	Agricultural employment		
Agro-processing industries	Agricultural expert		
	Immigration		

Source research findings and literature review

3 Results and Discussions

East Azerbaijan Province is the largest and most populous province in northwestern Iran. This province area is 45490.89 km^2 and is the 11th largest province of Iran. Weather conditions, soil characteristics, and geographical location of region have made the province suitable for agriculture. This province is one of the major agricultural centers of the country because of its fertile soil and extensive natural resources in agricultural activities. According to the latest political divisions in 2015, East Azerbaijan province has 21 counties (East Azerbaijan province governor 2015). Hurand County was separated from Kaleybar in 2013, but is not considered in this study due to the lack of sufficient information for this county. Therefore, this study is based only on information from 20 counties.

Table 2 summarizes the descriptive properties of the indicators for all three dimensions. The indicators have been normalized before being used in the weighting methods because they are measured in different units. Detailed information is available in the table and due to the large number of indicators no further explanation is given.

FAHP and entropy weighting results are reported in Table 3. The relative weights of agricultural sustainability indicators are calculated separately for each dimension. There is a great difference between results of two weighting methods. In economic dimension, insurance percentage and agro-processing industries with a big difference are the most important indicators of entropy method, while in FAHP method these indicators had the least relative weight. Also in social dimension, health indicator and

Table 2 Descriptive statistics of agricultural sustainability indicators

Dimension	Indicator	Units	Mean	Std.	Min	Max
Economic sustainability	Insurance percentage	%	17.31	14.49	2.18	51.718
	Irrigated wheat yield	Kg/ha	3902.64	697.121	2520.8	5096
	Efficiency of water consumption	Kg/m^3	3.864	2.344	1.048	11.915
	Economic participation rate	%	44.296	4.589	39.581	57.968
	Mechanization level	Hp/ha	1.592	0.363	1.09	2.55
	Agricultural land per capita	ha/person	5.24	2.359	2.578	11.060
	Conservation tillage	%	9.19	5.282	2.659	24.44
	Agro-processing industries	Numbers/1000 person	6.81	5.772	0	22.599
Social sustainability	Health indicator	%	93.5	4.383	86	100
	Population density	m^2/person	11841.8	6364.5	2968.9	26387.6
	Connection indicator	%	30.136	9.136	16.58	52.42
	Literacy rate	%	62.023	7.564	47.369	75.212
	Extension activities	Day-person/1000 person	80.463	38.736	16.694	174.292
	Agricultural cooperatives	Numbers/1000 person	2.189	1.304	0	5.485
	Agricultural employment	%	32.387	14.483	3.531	67.21
	Agricultural expert	Numbers/1000 person	3.535	1.156	2.424	7.42
	Immigration	–	1.053	0.659	0.236	3.271
Environmental sustainability	Groundwater level	m	1467.67	168.078	1191.46	1832.4
	Nitrate concentration	Mg/Lit	4.707	2.058	1.716	9.2
	Water salinity	μ mho/cm	1378.12	818.603	540.820	3461.72
	Efficient irrigation systems	%	4.721	5.534	0.641	23.510

(continued)

Table 2 (continued)

Dimension	Indicator	Units	Mean	Std.	Min	Max
	Agricultural water use	%	91.685	5.073	78.286	99.070
	Indicator of pesticide use	Kg/ha	0.783	0.976	0.0628	4.430
	Indicator of fertilizer use	Kg/ha	41.378	26.919	4.141	87.488
	Soil organic matter	%	1.389	0.264	0.928	1.834
	Fallow indicator	%	33.507	16.56	0.808	62.109
	Vegetation cover	%	45.29	14.66	23.744	80.745
	Grazing indicator	Animal unit (AU)/ha	7.906	4.643	2.407	16.909
	Percentage of greenhouse lands	%	0.927	1.373	0	5.044

Source Research findings

literacy rate got the most relative weight in FAHP method while these indicators got the lowest rank in entropy method. Altogether in economic dimension, the efficiency of water consumption and conservation tillage and in social dimension population density and agricultural employment indicators are the most important indicators and have a good rank in entropy and FAHP methods.

The experts identified the groundwater level, efficient irrigation systems, and nitrate concentration in groundwater as the most important indicators that influence the environmental sustainability. All these indicators show the important role of water resources sustainability in achieving the agricultural sustainability; but in the entropy method, the percentage of greenhouse lands, efficient irrigation systems, and pesticide usage indicators got the best ranks. These indicators have a non-uniform distribution, and there is great deal of variation among different counties of East Azerbaijan Province. In sum, the results of weighting estimation in environmental dimension show that the efficient irrigation system is the most important indicator in both methods.

Then we aggregate the weights of the decision elements to make a composite indicator for ranking the decision alternatives. Table 4 presents the results of agricultural sustainability ranking. It is expected that the ranking of counties is different in two weighting methods because the weight and rank of sustainability indicators were different in these two methods. First, we examine the entropy weighting results. Charoymaq, Hashtrood, Tabriz, and Jolfa Counties are the First Four, and Khoda Afarin, Kaleybar, Varzeghan, and Bonab Counties are the Last Four of East Azerbaijan Province in economic sustainability ranking. Increasing the percentage of agricultural insurance and Agro-processing industries has an important role in improving the economic sustainability of counties. The important role of agricultural insurance in sustainability is emphasized in previous studies like as Gomez-Limon

Table 3 The relative weights and ranks of agricultural sustainability indicators

Dimension	Indicator	Entropy		FAHP	
		Relative weight	Rank	Relative weight	Rank
Economic sustainability	Insurance percentage	0.308	1	0.0090	8
	Irrigated wheat yield	0.014	7	0.1080	3
	Efficiency of water consumption	0.136	3	0.5250	1
	Economic participation rate	0.005	8	0.0281	6
	Mechanization level	0.022	6	0.0396	4
	Agricultural land per capita	0.087	5	0.0332	5
	Conservation tillage	0.128	4	0.2440	2
	Agro-processing industries	0.300	2	0.0130	7
Social sustainability	Health indicator	0.001	9	0.3415	1
	Population density	0.169	3	0.0685	5
	Connection indicator	0.052	7	0.0132	7
	Literacy rate	0.009	8	0.2905	2
	Extension activities	0.144	4	0.0450	6
	Agricultural cooperatives	0.243	1	0.0126	8
	Agricultural employment	0.126	5	0.1343	3
	Agricultural expert	0.056	6	0.0820	4
	Immigration	0.200	2	0.0124	9
Environmental sustainability	Groundwater level	0.002	12	0.2202	1
	Nitrate concentration	0.036	9	0.1446	3
	Water salinity	0.038	8	0.1213	6
	Efficient irrigation systems	0.170	2	0.1747	2

(continued)

Table 3 (continued)

Dimension	Indicator	Entropy		FAHP	
		Relative weight	Rank	Relative weight	Rank
	Agricultural water use	0.056	6	0.1237	5
	Indicator of pesticide usage	0.165	3	0.0442	7
	Indicator of fertilizer use	0.141	4	0.0052	11
	Soil organic matter	0.006	11	0.0127	8
	Fallow indicator	0.048	7	0.0014	12
	Vegetation cover	0.015	10	0.1301	4
	Grazing indicator	0.058	5	0.0104	10
	Percentage of greenhouse lands	0.265	1	0.0115	9

Source Research findings

and Sanchez-Fernandez (2010) and Jamali Moghaddam et al. (2017). The first to the fifth rank of social sustainability belongs to Charoymaq, Heris, Hashtrood, Mianeh, and Jolfa Counties. Developing agricultural cooperatives, low population density, agricultural extension services, and high rate of agricultural employment are the common features of these counties. The findings of this study agree with the results of Koocheki et al. (2013) and Jamali Moghaddam et al. (2017) about agricultural cooperatives activities, Qian and Xueping (2007) and Ceyhan (2010) about population density and Gomez-Limon and Sanchez-Fernandez (2010) about agricultural employment. The results indicate that Marand, Malekan, and Khoda Afarin Counties receive the lowest social sustainability rank. Heris, Kaleybar, and Varzeghan are the most environmentally sustainable counties in East Azerbaijan Province while Malekan, Bonab, and Azarshahr are the least environmentally sustainable counties. The highest percentage of greenhouse cultivation, the lowest animal unit density in the rangelands or grazing indicator, and the lowest pesticide usage belong to Heris, Kaleybar and Varzeghan counties, respectively. The results of entropy method show that percentage of greenhouse cultivation, efficient irrigation systems and pesticide usage indicators receive the most relative weight among the environmental indicators and have an important role in environmental sustainability. These findings are according to the results of Pourzand and Bakhshode (2012) about pesticide usage.

The results of FAHP are relatively different from weights of entropy method because the basis of the work is distinct in these two weighting methods. According to FAHP results, the first to the fourth ranks of economic sustainability belong to Malekan, Ajabshir, Heris, and Bostan Abad counties, which have high water use efficiency consumption and good yield of irrigated wheat relative to other counties. The importance of crop yield also had been emphasized by the study of Radulescu et al.

Table 4 Sustainability ranking the counties of East Azerbaijan Province

	Economic sustainability		Social sustainability		Environmental sustainability		Total sustainability	
	Entropy	FAHP	Entropy	FAHP	Entropy	FAHP	Entropy	FAHP
Azarshahr	8	8	7	1	18	18	12	6
Osku	13	18	8	2	5	9	7	5
Ahar	11	7	10	17	11	10	13	17
Bostan Abad	9	4	12	19	9	1	8	4
Bonab	17	5	13	6	19	20	20	12
Tabriz	3	12	16	15	7	7	5	15
Jolfa	4	9	5	5	4	5	3	2
Charoymaq	1	11	1	7	17	12	4	13
Khoda Afarin	20	6	18	20	6	3	18	10
Sarab	12	19	11	12	15	6	15	16
Shabestar	14	14	17	4	10	16	16	8
Ajabshir	6	2	9	3	16	14	11	1
Kaleybar	19	17	6	18	2	2	6	11
Maragheh	16	15	14	16	14	13	17	20
Marand	5	20	20	8	12	15	14	19
Malekan	10	1	19	11	20	19	19	3
Mianeh	7	16	4	10	13	17	9	18
Varzeghan	18	13	15	13	3	4	10	7
Heris	15	3	2	14	1	8	1	9
Hashtrood	2	10	3	9	8	11	2	14

Source Research findings

(2010). Economically unsustainable counties of Marand, Sarab, Osku, and Kaleybar suffer from low efficiency of water consumption and low conservation tillage. These indicators had the most relative weights in FAHP method, and experts identified them as the most important economic indicators. The important effect of conservation practices on agricultural sustainability is also emphasized by the study of Dong et al. (2015). In social dimension, Azarshahr, Osku, Ajabshir, and Shabestar counties have the First Four ranks, and Khoda Afarin, Bostan Abad, Kaleybar, and Ahar counties have the Last Four sustainability ranks. The most important indicators of this dimension are health and literacy rate. These results are consistent with findings of Ceyhan (2010) and Jamali Moghaddam et al. (2017). Bostan Abad is the most environmentally sustainable county, which is followed by Kaleybar, Khoda Afarin, and Varzeghan Counties. High groundwater level, efficient irrigation systems, and low level of nitrate concentration are the common attributes of environmentally sustainable counties of East Azerbaijan Province in this method. Pourzand and Bakhshode (2012) have also used these indicators in their study, as the important factors of sustainability assessment.

According to the results of two weighting methods and ranking the counties in three considered dimensions, we do not expect that the ranking of total sustainability in two methods is similar. The total sustainability is obtained by the sum of the sustainability scores in three dimensions. Jolfa, Heris, Osku, Bostan Abad, and Ajabshir are ranked higher than other counties in both methods, and could be introduced as the most agriculturally sustainable counties of the province. However, Maragheh, Marand, Bonab, Sarab, and Ahar are the least agriculturally sustainable counties because of the low rank of total sustainability in both methods.

4 Conclusions and Policy Recommendations

Sustainability evaluation is a prerequisite for sustainable agricultural production and it involves assessment of the economic, social, and environmental parameters. The purpose of this study was to evaluate the agricultural sustainability of the East Azerbaijan Province counties using the composite indicator. The two employed methods were the Entropy of the positive weighting methods and fuzzy analytic hierarchy process of the exogenous weighting methods.

Our results enable us to identify the most important factors of agricultural sustainability in each dimension. It also helps us to rank the counties of East Azerbaijan Province in terms of agricultural sustainability. The results demonstrate that various weighting methods provide different results and solely relying on one method to evaluate the sustainability could lead to the wrong results. The ranking of the indicators and counties differs in the two methods used to weighting the indicators, but one can safely say that efficiency of water consumption and efficient irrigation systems indicators are among the most important agricultural sustainability indicators and Jolfa is among the most sustainable and Maragheh is among the most unsustainable counties in both methods.

The results of this study could help to improve current agricultural policies in East Azerbaijan Province, such as rural development policy, agricultural insurance policy, and water productivity policy in agricultural production, and modify the sustainability of the agriculture sector. Overall, based on the weighting results, to achieve higher levels of agricultural sustainability, it is recommended to increase the water use efficiency by changing the cropping pattern, accordance with the conditions of the region, and developing use of efficient irrigation systems (e.g., under pressure irrigation systems). Water is one of the key drivers of agricultural sustainability and there is an urgent need to increase the efficiency of its supply and make much better use of the applied water. Due to the drought crisis in the region, water resources sustainability is one of the main concerns of agriculture sustainability in East Azerbaijan Province, and local authorities and planners should pay serious attention to this matter. The health centers must also have more equitable distribution among the counties and more attention should be given to the health of people that live in deprived counties.

The results of this study showed that the extracted weights of sustainability indicators vary in the different methods, thus only relying on one method for extraction of the indicators' weights can lead to wrong results. Therefore, in order to accurate evaluation of agricultural sustainability in the future studies, it is suggested that do not trust the results of one method and also use the combination of the weighting methods.

References

Aggarwal R, Singh S (2013) AHP and extent fuzzy AHP approach for prioritization of performance measurement attributes. Int J Mech 7(1):6–11

Chang DY (1992) Extent analysis and synthetic decision, optimization techniques and applications. World Scientific Singapore

Ceyhan V (2010) Assessing the agricultural sustainability of conventional farming systems in Samsun province of Turkey. Afr J Agri Res

Ding L, Shao Z, Zhang H, Xu C, Wu D (2016) A comprehensive evaluation of urban sustainable development in China based on the TOPSIS-entropy method. Sustain J 8(746):1–23

Dong F, Mitchell PD, Colquhoun J (2015) Measuring farm sustainability using data envelope analysis with principal components: the case of Wisconsin cranberry. J Environ Manage 147:175–183

East Azerbaijan province governor (2015) Statistical yearbook Tehran Iran

Freudenberg M (2003) Composite indicators of country performance: a critical assessment. In: OECD science technology and industry working papers 2003/16 OECD Paris

Gomez-Limon JA, Sanchez-Fernandez G (2010) Empirical evaluation of agricultural sustainability using composite indicators. Ecol Econ 69:1062–1075

Hansen JW (1996) Is agricultural sustainability a useful concept? Agric Syst 50(1):117–143

Iravani H, Darban-Astaneh AR (2004) Measurement, analysis and exploitation of the sustainability of farming systems (case study: wheat production, Tehran province). Iran J Agric Sci 35(1):39–52

Jamali Moghaddam E, Yazdani S, Salami H, Peykani Gh (2017) Measurement of sustainability of farmers of Kamin plains in Fars Province: Comparison of PCA and AHP method. Iran J Agri Econ Dev Res 48(1):23–33

Jihad-Agriculture organization of East Azerbaijan province (2015) Statistical reports of agriculture jihad organization. Deputy of Planning and Economic Affairs

Koocheki A, Nassiri Mahallati M, Moradi R, Mansoori H (2013) Assessing sustainable agriculture development status in Iran and offering of sustainability approaches. J Agri Sci Sustain Prod 23(4):179–197

Kumar Singh R, Murty HR, Gupta SK, Dikshit AK (2010) An overview of sustainability assessment methodologies. Ecol Ind 9:189–212

OECD-JRC (Organization for Economic Co-operation and Development- Joint Research Centre) (2008) Handbook on constructing composite indicators: methodology and user guide. OECD Paris France

Pourzand F, Bakhshode M (2012) Evaluating agricultural sustainability of Fars province: application of compromise programming approach. J Agri Econ Res 4(1):1–26

Pretty JN (1995) Regenerating agriculture: policies and practice for sustainability and self-reliance. Earthscan Publications London

Qian L, Xueping H (2007) Assessment of the agricultural sustainability of Shaanxi Province China. J Ecol Econ 3:60–66
Radulescu CZ, Rahoveanu AT, Radulescu M (2010) A hybrid multi-criteria method for performance evaluation of Romanian South Muntenia Region in context of sustainable agriculture. In: Proceedings of the international conference on applied computer science, Malta, 15–17 Sept
Roy R, Weng Chan N, Rainis R (2014) Rice farming sustainability assessment in Bangladesh. Sustain Sci 9:31–44
Saaty TL (1980) The analytic hierarchy process. McGraw-Hill New York
Sauvenier X, Valckz J, van Cauwenbergh N (2006) Framework for assessing sustainability levels in Belgian agricultural systems—SAFE. Final Report—SPSD II CP 28. Belgian Science Policy Brussels
Shannon CE (1948) Mathematical theory of communication. Bell Sys Tech. J. 27:379–423
Statistical Center of Iran (2014) Iranian national census of agriculture. Islamic Republic of Iran Plan and Budget Organization
UNWCED (United Nations World Commission on Environment and Development) (1987) Our common future Brundtland report. Oxford University Press, Oxford UK

Marziyeh Manafi Mollayousefi is a lecturer and researcher at University of Tabriz, Iran. She holds a Ph.D. from the University of Tabriz (2017), and her research interest include agricultural and natural resource sustainability, natural resource economics, microeconomics, multi criteria decision making methods (MCDM) and sustainability evaluation. She has extensive teaching experience in the field of agricultural and natural resource sustainability, natural resource sustainability and microeconomics.

Babollah Hayati is a Professor in Department of Agricultural Economics at University of Tabriz. He was dean of Faculty of Agriculture during 2015–2018. He holds a B.Sc. in Agricultural Economics at University of Tehran and a M.Sc. and Ph.D. in Natural Resource Economics at Tarbiat Modares University. His areas of special interest are Natural Resource Economics, Sustainable Development Economics and Microeconomics. His recent publications have appeared in numerous journals including the Journal of Agricultural Science and Technology (JAST) and Engineering Sustainability.

Esmaeil Pishbahar is Associate Professor of Agricultural Economics at University of Tabriz, Iran. He holds a B.Sc. in Agricultural Economics from University of Tabriz and a M.Sc. in Agricultural Economics from University of Tehran. He did his Ph.D. in Science Economics at departments of Economics and Management, University of Rennes 1, France. His areas of interest and research are Applied Econometrics, Agricultural Risk Management and Insurance, and International Trade. His teaching area are Advanced Econometrics, Mathematical Economics, and Macroeconomics at under- and postgraduate levels. He has over 100 publications in journals and chapters in books.

Javad Nematian is an Associate Professor in Industrial Engineering Department at University of Tabriz, Iran. He holds a Ph.D. from Sharif University of Technology (2008). His research interests include uncertain programming, random Fuzzy & Fuzzy random optimization, combinatorial optimization problems, and mathematical programming.

Agricultural Producers and Consumers

Weather Risk Management: The Application of Vine Copula Approach

Sasan Torabi, Arash Dourandish, Mahmoud Daneshvar Kakhki, Ali Kianirad, and Hosein Mohammadi

Abstract Since unfavorable weather condition is the most important cause of loss in the gardening sector and apple production, it is necessary to provide a proper strategy for risk management to improve farm economic and social conditions. Insurance is a suitable tool to control such problems, while due to challenges, such as asymmetric information, it cannot manage these risks well. In this regard, it is logical to use the successful global approach in handling the weather-based index insurance, which can solve problems caused by asymmetric information, and can stabilize farmers' income. Thus, in this research, we design the weather-based index insurance for the apple production in Damavand which is considered an important apple production center in Iran. Data were collected during 1987–2016 from Iranian Agriculture Jihad Organization and the meteorological station in Damavand County. To investigate the dependency structure between weather variables and yield, the C- and D-vine copulas are used, and the Bayesian approach is employed to estimate the copula parameters. Considering the derived expected loss from this dependency structure, the premium is 36162340.44 Rials, that is different from the premium of the current insurance. This diversity arises form circular and administrative of current plan and lack of consideration of expected loss in the premium determination.

Keywords Weather-based index insurance · Canonical vine copula · Drawable vine copula · Apple · Dependency structure

---

S. Torabi · A. Dourandish · M. Daneshvar Kakhki (✉) · H. Mohammadi
Department of Agricultural Economics, Ferdowsi University of Mashhad, Mashhad, Iran
e-mail: daneshvar@um.ac.ir

S. Torabi
e-mail: sassan_torabi@yahoo.com

A. Dourandish
e-mail: dourandish@um.ac.ir

H. Mohammadi
e-mail: Hoseinmohammadi@yahoo.com

A. Kianirad
Agricultural Economics-Research, Deputy-APERDRI-Ministry of Agriculture, Tehran, Iran
e-mail: akianirad@gmail.com

M. Rashidghalam (ed.), *The Economics of Agriculture and Natural Resources*, Perspectives on Development in the Middle East and North Africa (MENA) Region,
https://doi.org/10.1007/978-981-15-5250-2_5

1 Introduction

Despite the importance of the agricultural sector, its production is inherently a risky business due to reliance on nature (Farzaneh et al. 2017). Nowadays, the number and severity of weather adverse events are exposed to increase. This issue has a negative effect on rural economic condition. Farmers greatly demand effective economic risk management tools that help them to cope with such events and to improve their economic conditions. Agricultural insurance is one of the effective policies in weather risk management. Even though insurance is an appropriate tool in risk management, it has problems, like asymmetric information that leads to higher premium rates and indemnity, and needs more accurate assessment of the claim to control the problems so that the insurers face surplus cost to evaluate the loss (Zhao et al. 2017). Such charge of traditional insurance schemes should be subsidized by the government. However, unfortunately, governments do not have sufficient funds to finance these subsidies at a large scale in developing countries. The experiences of developed and developing countries showed that by employing the index-based insurance, such as weather-based index insurance, we can solve a number of fundamental problems in traditional scheme (Conradt et al. 2015; Daron and Stainforth 2014; Jie et al. 2013; Bokusheva 2010; Leblois and Quirion 2010).

Unlike the traditional insurance plan, premium and indemnity are determined based on indices and their effect on loss of crops. Since the indices are clear and transparent, this insurance system can solve asymmetric information problems. Indemnities are not dependent on yield loss, and farmers endeavor hard to keep the products (Zhao et al. 2017). Also, using the factors which are out of farmers' control reduces the occurrence probability of moral hazard and adverse selection. In addition, unlike traditional crop insurance, the insurer does not need to visit a farmer's field to determine premiums and assess damages. If the value of weather index is below or above the agreed threshold, then the insurance pays out (Aziznasiri 2011). Considering the benefits of index-based insurance and the global tendency toward this type of insurance, it is expected that the Agricultural Insurance Fund apply this system in Iran.

There are numerous studies about weather-based index insurance: For instance, Zhao et al. (2017) pointed that adverse selection is an inappropriate behavior of farmers in China so that it leads to ruin and bankrupt the current system of agricultural insurance. Ma and Maystadt (2017) examined the effects of weather variables on corn yield and farmers' income in China. For this, the information of 4861 farmers was collected between 2004 and 2010. The estimation results of the panel model showed that increasing temperature and drought had a negative effect on corn yield and farmers' income. Another group of studies identified the effects of weather factors on crop yield. Burke et al. (2015) showed that high temperature had a profound negative impact on crop yield. Since 1981, global warming has resulted in annual combined losses of three crops (wheat, maize, and barley) about 5 billion dollars per year (Lobell and Field 2007). Experimental evidence in Africa also showed that weather variables had a significant effect on grain yield (Schlenker and Lobell 2010;

Lobell et al. 2011; Roudier et al. 2011; Blanc 2012). Dell et al. (2014) showed that the high temperature has a negative effect on rice yield in India, and the same results were also found in Indonesia, the Philippines, and Thailand (Guiteras 2007; Welch et al. 2010). Tao and Zhang (2010), Zhang and Huang (2012), Yao et al. (2014), and Zhang et al. (2015) proved that temperature plays an important role in the corn yield. In Iran, some studies investigated and designed the weather-based index insurance in different parts of the country. For example, Kochakzaei and Kochakzaei (2015) showed that weather-based index insurance is a risk transfer tool for agricultural production in low-income countries. Researchers found that this system is the main requirement and effective mechanism to transfer the natural disasters risks, and to increase farmers' income. Pishbahar et al. (2015) studied the dependency structure between wheat yield and weather variables and calculated the premium. The results showed that this system's premium was lower than the current premium. Furthermore, according to dependency structure, the relative humidity had the highest correlation with wheat yield. Kochakzaei et al. (2013) determined the appropriate correlation model, and showed that the correlation of wheat yield and rainfall index was up to 98 percent. Then, they calculated the yield reduction caused by rainfall reduction and the rainfed wheat premium using this index. They concluded that the use of weather-based index insurance could protect the agricultural sector against adverse natural factors, and increase the efficiency of crop insurance policy. In Iran, the gardening sub-sector is one of the most important agricultural and economical parts. According to Food and Agricultural Organization (FAO) (2014), the value of this sub-sector production was equivalent to 22% of the total value of the agricultural production. Apple, as the main product in this sub-sector, is stated the first among the gardening products with a 48.3 million tons production, and is accounted for 18% of the total production of gardening sub-sector. Also, in recent years apple trees are commercially cultivated and the products are exported to global markets. The high yield level and commerciality of apple product have encouraged the farmers to make new apple gardens. The area planted with apple trees is about 208.5 thousand ha, which is 9.1% of the total horticultural land area. Tehran Province, with 10% of apple production, is one of the important apple producers of the country, in which total production and yield are 362523.66 and 33.5 tons per hectare, respectively (Agricultural Statistics 2015). In this province, Damavand ranked first with 223 thousand tons of production. Damavand is a county in the east of Tehran Province and its capital is Damavand. This county is divided into two districts: the central and Rudehen. Damavand County has five cities, Damavand, Abali, Rudehen, Absard, and Kilan. According to 2011 census, the county's population is about 757,500 in 27,419 families. Damavand has a mountain and cold climate (Governor's Office of Damavand 2015). This region is also one of the apple production centers in the country, which accounts for 6.4% of the country's total production (Iranian Agricultural Organization site, Tehran Province 2015).

Horticultural products, like apple, are exposed to variety of risks, such as cold, frost, drought, and hail. In 2014–2015 crop year, Agricultural Insurance Fund reported 583,203 damages in the horticulture sub-sector, in which 82% of them were caused by weather hazards. In recent years, the loss of apple caused by inappropriate weather condition in the country and Tehran Province was notable so that the loss was equal to 10% of the total horticultural products' loss. In Tehran, especially in Damavand, frostbite is an important cause of loss to apple, and frost and cold lead to 70% loss. Accordingly, it is perspicuous that the main reason for apple loss is due to the inappropriate weather situation and climate changes; hence, it is essential to apply a suitable policy to control weather risk in Damavand and country.

Regarding the benefits of index-based insurance and the global motion toward this type of insurance, it is expected that the Agricultural Insurance Fund employs this insurance plan for agricultural crops. Thus, with attention to challenges of the traditional agricultural insurance and the numerous benefits of weather-based index insurance, this study attempts to design the weather-based index insurance for Damavand apple.

Our paper is organized as follows. We present the methodology in Sect. 2, and afterward in Sect. 3 applies the method to the Iranian case study. Finally, Sect. 4 gives conclusion and policy recommendations.

2 Methodology

We need to calculate the fair premium to design an insurance system. In the literature, the premium calculation methods are classified into two main groups. The first so-called expected utility method considers farmers' risky behavior in the decision-making process. In contrast, the second method merely examines the behavior of yield and risk factors. Then, according to the statistical methods, it calculates the dependency structure and expected loss to determine the premium (Robison and Barry 1987). Because of the limitations of expected utility in considering all risky behaviors in the decision-making process, the statistical ways are usually applied. Therefore, in this study, we use the second method to determine the premium. In statistical methods, we investigate the dependency structure between yield and weather variables that cause stress for the crop. Then, based on the dependency structure, we simulate the conditional yield and compute expected loss. Therefore, it is essential to determine the appropriate method to derive the dependency structure. The percentage of non-linear risks has increased in risk management and insurance. This issue attracted researchers' attention to non-Gaussian models in dependency structure determination. In this regard, searching for flexible multivariate distributions, copula approach became instrumental in many fields of application (Brechmann et al. 2010).

The copula is a grounded function that connects the marginal distributions to generate a multivariate distribution (Chen et al. 2013). In other words, let $X = (x_1, \ldots, x_d)'$ be a d-dimensional vector of random variables with joint distribution $F = F(x_1, \ldots, x_d)$ and margins $F_1(x_1), \ldots, F_d(x_d)$. Then, there is a copula C which transforms the marginal distributions to joint distribution using their dependency structure (Kim et al. 2013). According to Sklar theorem (1959), a copula is a multivariate distribution function on $[0,1]^d$ with uniform marginal distributions,

$$F(x_1, \ldots, x_d) = C(F_1(x_1), \ldots, F_d(x_d); \theta) \tag{1}$$

where θ is the parameter of copula that shows the dependency between the marginal distributions (Geidosch and Fischer 2016). In general, the copulas are categorized into implicit and explicit. Implicit or Elliptical copulas like Gaussian and Student's t are extracted from a certain theoretical distribution functions, and they have only symmetrical tail dependency. While explicit copulas, i.e., Archimedean and Extreme value, are obtained from the generator and the dependence function, respectively. The most known examples for Archimedean copulas are Clayton, Gambel, Joe, and Frank, and for Extreme value copulas are Tawn, Gambel, Galambos and Husler and Reiss (Beatriz et al. 2005).

The simple copulas are restricted in an arbitrary dimension. A large number of variables often have complex dependence model (Brechmann and Schepsmeier 2012). Therefore, high dimensional data requires flexible multivariate models that can provide a suitable dependency structure. One of these models is the vine copula that was introduced initially by Joe (1996) and was developed by Bedford and Cooke (2001, 2002) and Kurowicka and Cooke (2006). Vine copula is a flexible graphical or matrix model to explain joint distributions built up using a group of bivariate copulas as the so-called "pair copula constructions" (PCC). The PCC is a very useful way to create a flexible multivariable or joint distribution. This model decomposes the multivariate distribution into a cascade of bivariate copula as a tree structure. Aas et al. (2009) showed that these models could be expanded with optional families of pair copulas. Thus, a multivariate distribution, which is decomposed into pair copulas from different types of families, is called "mixed vine."

Let $X = (X_1, \ldots, X_d)^t$ with a joint density of $f(x_1, \ldots, x_d)$ and marginal densities of $f_1(x_1), \ldots, f_d(x_d)$. $f(x_1, \ldots, x_d)$ to be decomposed as

$$f(x_1, x_2, \ldots, x_d) = f_d(x_d).f(x_{d-1}|x_d).f(x_{d-2}|x_{d-1}, x_d) \ldots f(x_1|x_2, \ldots, x_d) \tag{2}$$

It can be shown that the joint density function $f(x_1, \ldots, x_d)$ for an absolute and continuous multivariate distribution F with strictly increasing continuous marginal densities is

$$f(x_1, x_2, \ldots, x_d) = c_{1,\ldots,d}(F_1(x_1), \ldots, F_d(x_d)).\prod_{i=1}^{d} f_i(x_i) \tag{3}$$

where $c_{1,\ldots,d}$ is a d-variate copula density. A bivariate joint distribution is represented as a product of a copula and marginal densities as

$$f(x_1, x_2) = c_{1,2}(F_1(x_1), F_2(x_2)) \cdot f_1(x_1) \cdot f_2(x_2) \tag{4}$$

In addition, the conditional density can be presented in terms of a copula as follows:

$$
\begin{aligned}
f(x_2|x_1) =& \frac{f(x_1, x_2)}{f_1(x_1)} = \frac{c_{12}(F_1(x_1), F_2(x_2)).f_1(x_1).f_2(x_2)}{f_1(x_1)} \\
=& c_{12}(F_1(x_1), F_2(x_2)).f_2(x_2)
\end{aligned}
\tag{5}
$$

Equation (5) in the d-dimensional case is as follows:

$$
f(x|\upsilon) = c_{x\upsilon_j|\upsilon_{-j}}(F(x|\upsilon_{-j}), F(\upsilon_j|\upsilon_{-j})) \cdot f(x|\upsilon_{-j})
\tag{6}
$$

where υ denotes a d-dimensional vector, υ_j shows an arbitrary selected element of υ, and υ_{-j} is a vector of υ, excluding the arbitrary element. In the 3-dimensional case, the density function is

$$
\begin{aligned}
f(x_1, x_2, x_3) =& f_1(x_1).f(x_2|x_1).f(x_3|x_1, x_2) \\
=& f_1(x_1). \underbrace{f_2(x_2) \cdot c_{12}(F_1(x_1), F_2(x_2)}_{f(x_2|x_1)} \cdot \\
& \underbrace{f_3(x_3) \cdot c_{13}(F_1(x_1), F_3(x_3)) \cdot c_{23|1}(F(x_2|x_1), F(x_3|x_1))}_{f(x_3|x_1,x_2)}
\end{aligned}
\tag{7}
$$

The PCC consist of marginal conditional distributions $F(x|\upsilon)$. Joe (1996) indicated that for every u_j in the vector υ, $F(x|\upsilon)$ can be written as follows:

$$
h(x|\upsilon, \theta) = F(x|\upsilon) = \frac{\partial C_{x\upsilon_j|\upsilon_{-j}}(F(x|\upsilon_{-j}), F(\upsilon_j|\upsilon_{-j})|\theta)}{\partial F(\upsilon_j|\upsilon_{-j})}
\tag{8}
$$

where $C_{x\upsilon_j|\upsilon_{-j}}$ is an optional copula function that can be selected from different families (Emmanouil and Nikos 2012). It is clear that the decomposition of Eq. (2) can be done in different ways so that there are a remarkable number of PCCs in high dimensions. Therefore, Bedford and Cooke (2001, 2002) presented the graphical model, known as "regular vine copula" (R-vine). R-vine copulas include two main groups, including Canonical vine copula (C-vine) and Drawable vine copula (D-vine). The main difference between them is the decomposition form of the multivariate density function. The C-vine tree has a star structure, while the D-vine tree has a path structure.

As mentioned above, the decomposition of the joint density with a C-vine pattern leads to a set of trees with star form. The pairs corresponding to (l,i), $i = 2,\ldots,d$, in the first tree of a d-dimensional C-vine copula, and for tree l, $2 \leq l < d-1$, the pairs are specified as $(l, i|1, \ldots, l-1), i = l+1, \ldots, d$. Figure 1 shows a 5-dimensional example of the C-vine. In this figure, there is a node (or variable) with the maximal degree in each tree that is connected to the other nodes and called the central node. The node that makes the strongest dependency in each tree is selected as the central node. The central node is also known as the pilot node.

A C-vine density is as follows (Aas et al. 2009):

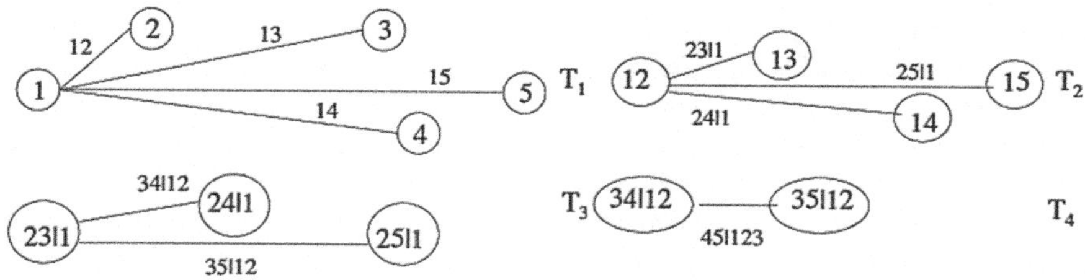

Fig. 1 Example of a 5- dimensional C-vine copula's trees

$$f(x) = \prod_{k=1}^{d} f_k(x_k) \prod_{i=1}^{d-1} \prod_{j=1}^{d-i} c_{i,i+j|1:(i-1)} \left(F(x_i|x_1, \ldots, x_{i-1}, F(x_{j+1}|x_1, \ldots, x_{i-1})|\theta_{i,i+j|1:(i-1)}\right) \tag{9}$$

In a D-vine, in the first tree, the dependence of the first and second node, of the second and third, and so on, is modeled using an edge (or pair copula). In other words, in the first tree the pairs correspond to $(i,i + 1)$, $i = 1,\ldots, d$, and in the second tree, the conditional dependence is modeled as $(i, i + 2|i + 1)$. Consequently, the dependencies of variables are modeled in tree $T_j, j = 1,\ldots, d-1$, conditioned on those variables that lie between the variables in tree T_{j-1}. Figure 2 shows a 5-dimensional example of D-vine. According to this figure, no nodes are connected to more than two edges in each tree. Finally, the connection of variables makes a path structure.

A D-vine density is given as follows (Brechmann and Schepsmeier 2012):

$$f(x) = \prod_{k=1}^{d} f_k(x_k) \prod_{i=1}^{d-1} \prod_{j=1}^{d-i} c_{j,j+i|(j+1):(j+i-1)} \left(F(x_j|x_{j+1}, \ldots, x_{j+i-1}, F(x_{j+1}|x_{j+1}, \ldots, x_{j+i-1})|\theta_{j,j+i|(j+1):(j+i-1)}\right) \tag{10}$$

Figures 1 and 2 contain four trees. Tree T_j has $d + 1-j$ nodes and $d-j$ edges. Each edge corresponds to a pair copula density. For example, in Fig. 1, the edge (34|12) corresponds to the conditional copula density, $c_{34|12(0)}$. The edges of tree T_j

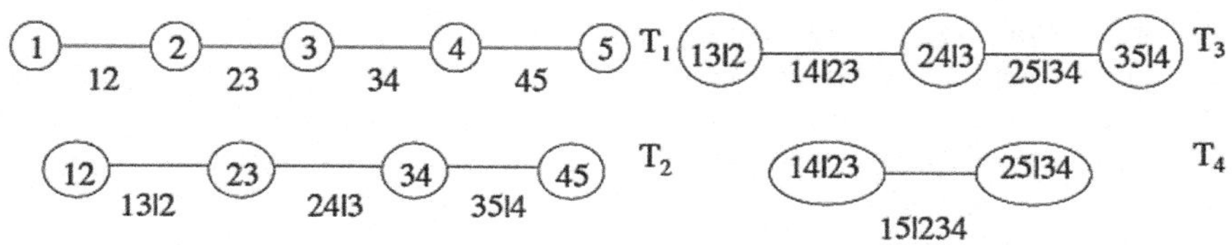

Fig. 2 Example of a 5- dimensional D-vine copula's trees

become the nodes of tree T_{j+1}. There are as many as $d!/2$ different PCCs for C-vine and D-vine. Following Aas et al. (2009), we select the tree structure that has the strongest dependencies based on the maximum sum of absolute Kendall's tau in each tree. It is also necessary to choose $d(d-1)/2$ pair copulas for each structure; for this, we use AIC and BIC statistics. The copula data are in [0,1]; however, the general data do not have uniform margins. To transform the data, we use the empirical cumulative distribution function. Another critical point is the parameter estimation. The agricultural data are limited, and this issue might affect the validity of models such as, the maximum likelihood. To cope with this problem, following Bokusheva (2010), we apply the Bayesian method to estimate the pair copulas parameter.

The Bayesian method considers the parameters, θ, as random variables and focuses on the estimation of parameters entire distribution than point estimation. In this method, a distribution is assumed for the parameters that are called prior distribution ($\pi(\theta)$), which incorporates the prior information about the distribution of parameters. It does not depend on the observed data. Regarding the observations information, x, the prior distribution is adjusted, and new distribution for the parameter is generated that is called the posterior distribution ($P(\theta) = p(\theta|x)$). The posterior distribution can be represented by the product of the likelihood function, $L(\theta|x)$, and the prior density function, $\pi(\theta)$, as (Czado et al. 2014):

$$p(\theta|x) = \frac{L(\theta|x) \cdot \pi(\theta)}{f(x)} \propto L(\theta|x) \cdot \pi(\theta) \tag{11}$$

The likelihood function for each pair copula can be represented as follows:

$$L(\theta|x) = \prod c(F_{1_1}(x_1|\theta), F_2(x_2|\theta)) \cdot f_{1_1}(x_1|\theta) \cdot f_2(x_2|\theta) \tag{12}$$

Finally, the Bayesian estimation of the parameter is obtained as

$$\hat{\theta}_B = \int \theta p(\theta|x) d\theta = \int \theta \pi(\theta) L(\theta|x) d\theta \tag{13}$$

The calculation of $\hat{\theta}_B$ is not simple, and we use the approximation of $\hat{\theta}_B$. The Markov chain Monte Carlo (MCMC) technique is a rational method to approximate the parameters' posterior distribution. In this method, we make the Markov chain by sequential sampling from entire conditional distributions of parameter using the Metropolis–Hastings algorithm. Upon convergence, its states denote draw from the desired distributions, and based on, we can infer about the model parameters. The Metropolis–Hastings algorithm actually implements a simulation of a Markov chain through a mechanism of acceptance/rejection. This algorithm generates $\theta^{(t+1)}$ by $\theta^{(t)}$ and the transmission probability from the state t to the state $t+1$. For this purpose, an arbitrarily initial value of $\theta^{(0)}$ with an optional distribution, $\Pi(0)$, and selection of the transferring probability core, $q(0|\theta)$, are required to obtain other observations (Carlin and Louis 2000).

When the dependency structure is determined by proper vine copula, we simulate the yield concerning the weather variables to calculate the expected loss. The Monte Carlo conditional simulation method is one of the common methods for simulating the data. In this method for each dimension, we select a sample and then generate data using the conditional distribution function inverse with respect to the selected variable. The simulated data are copula data as well. We apply the "inverse cumulative distribution function" to convert them into their real form. Thus, we need to determine the distribution of the variables. Thus, statistics, like "Kolmogorov–Smirnov," "Anderson–Darling," and "Chi-Squared" can be used.

We predict the yield using the ARIMA process. Using the yield predicted value, critical value of the yield at three coverage levels (50, 70, and 100%) is $y_C = y_f \times \text{COV}$. Where y_f is the predicted yield, and COV is the coverage level. Next, we compare simulated yields (y_i) with y_C. In case each of them is less than critical level, the insurer should pay the indemnity. The expected loss is equal to $\text{Ave}[\max(y_c - y_i, 0)]$. Therefore, the fair premium is calculated by $\text{Ave}[\max(y_c - y_i, 0)] \times P$, where P indicates the pre-determined price by the Agricultural Insurance Fund. The insurer always adds an amount to the fair premium because of the implementation costs of insurance. This amount which is called "loading factor" is known as a percentage of fair premium. According to Pishbahar et al. (2015), the implementation costs is about 10% of fair premium, hence, the actual premium can be calculated by dividing the fair premium to 0.9 (Pishbahar et al. 2015, 2019).

3 Results and Discussion

In this study, we use the following weather variables: weighted average temperature (T), cumulative rainfall index (CRI), relative humidity (RH) in phenological stages, and faster wind speed (U) in apple ripening time. These variables weight represents a sub-period's weight which is obtained from standard regressions of weather variables on apple yield (Y). The apple yield and weather variables' information were collected over the period from 1987 to 2016 from the Iranian Agricultural Organization and meteorological station in Damavand County. The data description result is reported in Table 1. The average yield of apple is about 30.8 tons/ha, and the minimum and the maximum are 23.2 and 37 tons, respectively. The minimum temperature

Table 1 The results of data description

Variable	Mean	Sd	Max	Min
Y (ton)	30.8	8.39	37	23.2
T (°C)	12.98	1.49	33.98	−8.62
CRI (mm)	331.79	27.94	11.39	168.46
RH (%)	45.59	6.75	72.59	25.47
U (knot)	12.03	2.79	18	8

associated with hibernation stage is −8.62 °C and the maximum temperature is 33.98 °C in growth stage. The average temperature of the region is 12.98 °C, which is classified as a cold and mountainous climate. The cumulative rainfall during the phenological stages is 331.79 mm, and the maximum amount of rainfall is 168.46 mm in hibernation stage and the minimum is 11.39 mm in ripening stage. The average relative humidity at different phenological stages is 45.59% and the average fasten wind speed in the fruit ripening stage is 12.03 knots.

As mentioned above, in order to determine the premium, we must calculate the expected loss and as a result simulate the yield considering the weather variables that cause stress on crop yield. We investigate the dependency structure using various types of vine copulas to simulate the variables. The copula data are $Y_{\text{ECDF}} = 1$, $U_{\text{ECDF}} = 2$, $T_{\text{ECDF}} = 3$, $\text{CRI}_{\text{ECDF}} = 4$, and $\text{RH}_{\text{ECDF}} = 5$. The variables 1, 2, 3, 4, and 5 denote yield, faster wind speed, temperature, cumulative rainfall index, and relative humidity, respectively. In fact, these five copula data are the nodes' set in the first tree, ($N_I = \{1, 2, 3, 4, 5\}$). In C-vine structure the central node (pilot node) in each tree is determined. For this purpose, we calculate the Kendall's tau for all pairs of variables $\{j, k\}$. A node with the maximum sum of absolute Kendall's tau is the pilot node. Then, according to minimum value of AIC and BIC, we select the suitable pair copula for each edge between pilot node and others, and estimate the parameters using Bayesian approach in the first tree (Table 2).

According to Table 2, the sum of absolute Kendall's tau (or the spanning tree) indicates that the node 5 or ECDF of relative humidity has the highest sum and is the pilot node in the first tree. Therefore, the edges' set is as $E_1 = \{(5, 1), (5, 2), (5, 3), (5, 4)\}$ in the first tree, and selected copulas for each of them are Clayton, Gaussian, Joe 180, and Frank, respectively. The estimated parameters using the Bayesian approach

Table 2 Results of Kendall's tau, edges, copula families and their estimated parameters in the first tree

Nodes	Kendall's tau				
	1	2	3	4	5
1	–	−0.07	0.54	0.30	0.36
2	−0.07	–	−0.04	0.15	0.13
3	0.54	−0.04	–	0.35	0.46
4	0.30	0.15	0.35	–	0.45
5	0.36	0.13	0.46	0.45	–
Sum	1.28	0.39	1.40	1.26	1.41
Pair copulas (edges)	E_1 =	{(5,1),	(5,2),	(5,3),	(5,4)}
Selected copulas	Family_1 =	{Clayton,	Gaussian,	Joe 180,	Frank}
Parameter posterior distribution mean	Par =	0.91	0.22	2.81	4.73
Standard deviation	Sd =	0.34	0.09	0.50	1.33
Tail dependency (upper, lower)		(0,0.47)	(0,0)	(0,0.72)	(0,0)

Table 3 Results of Kendall's tau, edges, copula families and their estimated parameters in the second tree

Nodes	Kendall's tau			
	1,5	2,5	3,5	4,5
1,5	–	−0.15	0.39	0.21
2,5	−0.15	–	−0.14	0.02
3,5	0.39	−0.14	–	0.20
4,5	0.21	0.02	0.20	–
Sum	0.76	0.31	0.73	0.43
Pair copulas (edges)	E_2 =	{(1,2|5)	(1,3|5),	(1,4|5)}
Selected copulas	Family_2 =	{Gaussian,	Frank,	Frank}
Parameter posterior distribution mean	Par =	−0.14	4.84	1.98
Standard deviation	Sd =	0.09	1.34	1.13
Tail dependency (upper, lower)		(0,0)	(0,0)	(0,0)

show that the variables are not independent and do not have a specific economic interpretation (Pishbahar et al. 2015). The Clayton and the Joe180 exhibit strong lower tail dependence, while the Gaussian and Frank have a symmetric dependency structure.

The edges in the first tree become nodes in the second tree. To select the pilot node it is necessary to create input. For this, we form pseudo-observations according to Eq. (8). The nodes set in the second tree are $N_2 = \{(5,1), (5,2), (5,3), (5,4)\}$ that shows we have four representatives for the second pilot node. Similarly, we calculate the Kendall's tau for newly created input. The results of the second tree are presented in Table 3.

The result of spanning tree in Table 3 implies that the node (5,1) has the maximum sum of absolute Kendall's tau and is the second pilot node. In this regard, the edges set is $E_2 = \{(1,2|5), (1,3|5), (1,4|5)\}$ so that the pilot node in the first tree or 5 becomes conditioning element. Gaussian, Frank, and Frank copulas are selected for these edges, which represent a symmetric dependency structure in the second tree. The estimated parameters also indicate the relationship between nodes. Likewise, third and fourth trees are also created and the results are given in Tables 4 and 5, respectively.

According to Tables 4 and 5, the third pilot node is (1,2|5). Therefore, the order of copula data in this C-vine model is 5, 1, 2, 3, and 4, respectively. There is also symmetric dependency structure in these two trees. It should be noted that the plots for posterior distribution of copula parameters are displayed in Appendix. Thus, by determining the rank of nodes in the C-vine pattern, the tree structure is obtained (Fig. 3). This tree structure actually represents the joint density function of the yield and the weather variables as a set of nodes of marginal distributions in the first tree, the non-conditional pair copulas in the second tree and the conditional pair copulas in the third and fourth tree.

Table 4 Results of Kendall's tau, edges, copula families and their estimated parameters in the third tree

Nodes	Kendall's tau		
	1,2|5	1,3|5	1,4|5
1,2|5	–	−0.06	0.07
1,3|5	−0.06	–	0.05
1,4|5	0.07	0.05	–
Sum	0.13	0.11	0.12
Pair copulas (edges)	E_3 =	{(2,3|5,1),	(2,4|5,1)}
Selected copulas	Family_3 =	{Frank,	Gaussian}
Parameter posterior distribution mean	Par =	−0.81	0.03
Standard deviation	Sd =	1.08	0.1
Tail dependency (upper, lower)		(0,0)	(0,0)

Table 5 Results of Kendall's tau, edges, copula families and their estimated parameters in the fourth tree

Nodes	Kendall's tau	
	2,3|5,1	2,4|5,1
2,3|5,1	–	0.04
2,4|5,1	0.04	–
Sum	0.04	0.04
Pair copulas (edges)	E_4 =	{(3,4|5,1,2)}
Selected copulas	Family_4 =	{Gaussian}
Parameter posterior distribution mean	Par =	0.12
Standard deviation	Sd =	0.09
Tail dependency (upper, lower)		(0,0)

As mentioned above, in the D-vine tree structure, we must determine the order of nodes in the first tree. The structure of this vine copula is like a path, so-called Hamiltonian path. In this model, maximum Hamiltonian path should be identified. This issue is like a traveling salesman problem (TSP). In this case, as a Nondeterministic Polynomial problem, there is no definite algorithm to find a solution (Czado et al. 2014). Using this algorithm and based on the maximum sum of absolute Kendall's tau, the order of variables in the D-vine tree is determined. Since there are five variables in the model, we need to compare $\frac{5!}{2} = 60$ different kinds of D-vine copulas. For this purpose, we use the presented results of the Kendall's tau in Table 2. By comparing 60 types and applying the traveling salesman algorithm, the best order of variables are 1, 3, 5, 4, and 2, respectively. Thus, in the first tree, the spanning tree is equal to 1.6.

Afterward, based on the minimum value of AIC and BIC, the proper copulas for each edge of the paths in the tree set are selected, and the corresponding parameters

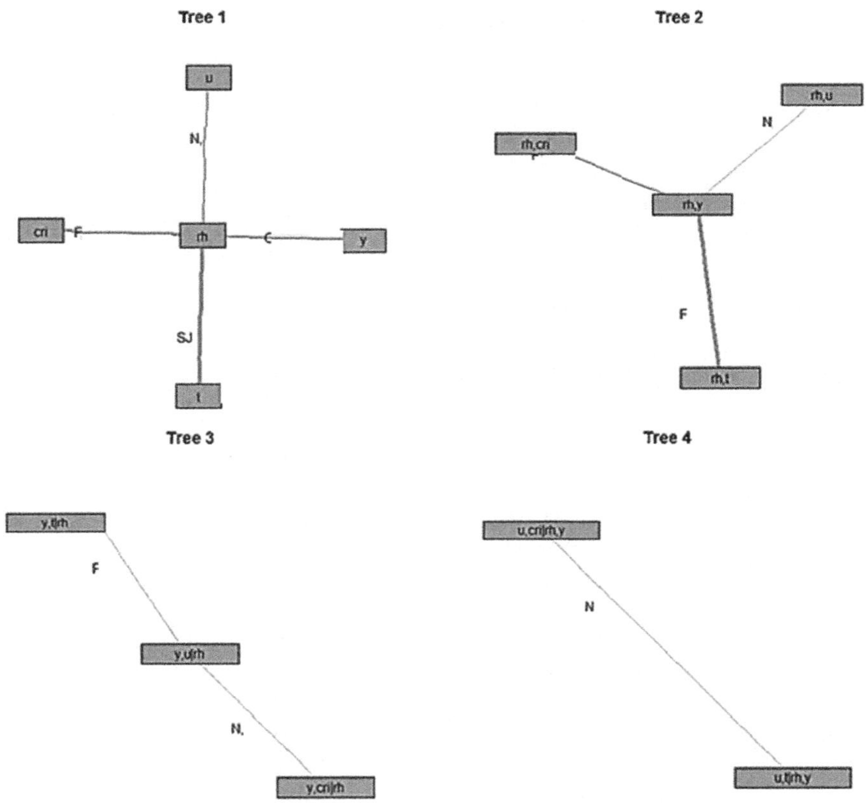

Fig. 3 Representation of C-vine tree structure for apple yield and weather variables

are estimated using Bayesian approach. Here, we considered normal distribution as a prior distribution for parameters. The results of copula selection and their parameter estimation are also presented in Table 6, and the posterior distribution diagrams are reported in Appendix.

Insert Table 6 about here it should be noted that some copulas have different number of parameters (one, two, or three parameters), that according to Table 6 the Tawn I and II copulas have two parameters. Unlike the C-vine copula in the D-vine copula, most of the copulas are selected from extreme value families. Estimated parameters show that conditional and non-conditional pairs are not independent. Also, except for Clayton, Joe 180, and TawnII 180, which exhibit lower tail dependence, the rest represent a symmetric dependency structure between the variables. The results of the two copulas (3.5) and (5.4) are identical in the first tree of C-vine and D-vine, because this relationship is on the first level and does not represent the overall structure of the trees. Thus, by determining the order of variables in the D-vine structure and the selection of copulas and estimation of their corresponding parameters, the

Table 6 Results of copula families and their estimated parameters in D-vine

Trees	Edges	Copulas	Parameter posterior distribution mean	Standard deviation	Tail dependency (upper, lower)
First	1,3	Clayton	2.05	0.51	(0,0.71)
	3,5	Joe180	2.81	0.49	(0,0.72)
	5,4	Frank	4.73	1.35	(0,0)
	4,2	Frank	1.27	1.19	(0,0)
Second	1,5|3	P_1 of TawnI 270	−38.50	32.39	(0,0)
		P_2 of TawnI 270	0.05	0.02	
	3,4|5	Gaussian	0.17	0.09	(0,0)
	5,2|4	P_1 of TawnII 180	46.93	35.11	(0,0.069)
		P_2 of TawnII 180	0.07	0.02	
Third	1,4|5,3	P_1 of TawnII 180	53.95	34.39	(0,0.02)
		P_2 of TawnII 180	0.06	0.01	
	3,2|4,5	Frank	−0.98	1.06	(0,0)
Forth	1,2|5,3,4	P_1 of TawnI 270	−55.53	41.15	(0,0)
		P_2 of TawnI 270	0.06	0.04	

tree structure of this vine copula is obtained and is presented in Fig. 4. This structure also shows the joint density function of the apple yield and weather variables in the term of a D-vine pattern.

When the tree structure is determined, two patterns are compared to define the best one in joint distribution explanation. For this, we use the AIC, BIC and Vuong and Clark tests (Table 7). In Vuong test, the null-hypothesis shows D-vine is better, but Clarke test shows that two models are statistically equivalent. The results indicate that the null-hypothesis is rejected in both tests. In addition, AIC and BIC also show that the C-vine model is better than the D-vine model in explaining the joint distribution between Damavand apple and weather variables.

In addition, we determine the proper R-vine matrix structure for mentioned variables that its result is as follows:

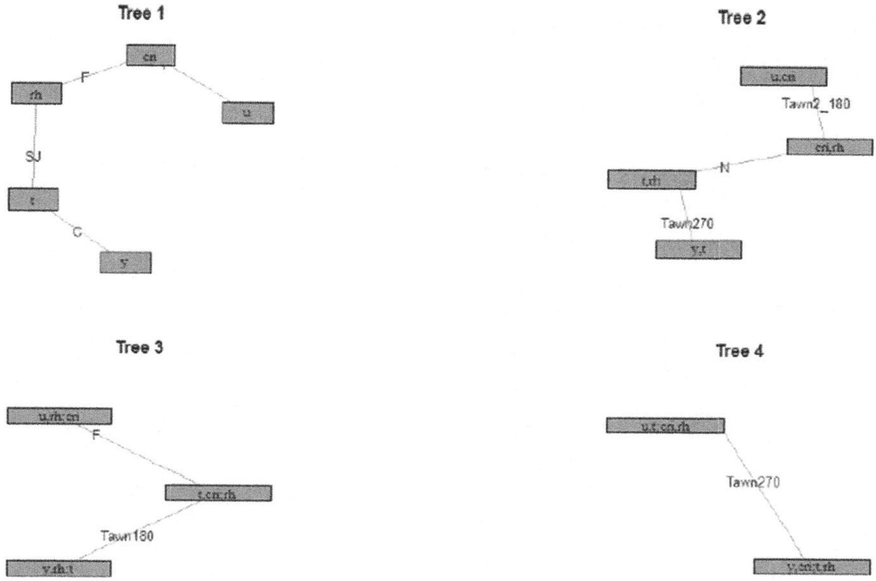

Fig. 4 Representation of D-vine tree structure for apple yield and weather variables

Table 7 Results of vine copula selection

Tests		Statistic	*P*-value		
Clark		15	0.01		
Vuong		1.62	0.04		
C-vine	AIC	−49.67	D-vine	AIC	−64.31
	BIC	−35.65		BIC	−44.69

$$\begin{bmatrix} 4 & 2 & & & \\ 2 & 4 & 1 & & \\ 1 & 1 & 4 & 4 & \\ 5 & 5 & 5 & 5 & 5 \end{bmatrix}$$

Col1	Col2	Col3	Col4
3,4|2,1,5	2,4|1,5	1,4|5	4,5
3,2|1,5	2,1|5	1,5	
3,1|5 5	2,5 5 5	5	5
3,5			

Regarding the R-vine matrix structure, it is clear that this pattern matches the C-vine pattern, and this is another evidence of suitability of the C-vine copula in explaining the joint distribution of apple yield and weather variables. Therefore, we use the results of this vine copula to simulate the yield variable concerning weather variables. The simulated data are also copula data that are converted to their real appearance using inverse cumulative distribution of yield. To select the appropriate distribution for yield, the three mentioned tests are used and the results are reported in Table 8. According to the minimum amount of three mentioned tests,

Table 8 Theoretical distribution selection for Damavand apple yield

Selected distribution		Kolmogorov-Smirnov	Anderson-Darling	Chi-squared
Wakeby	Statistic	0.088	0.374	0.999
	P-value	0.957	–	0.801
	Critical value (α = %5)	0.241	2.501	7.814

Table 9 Premium in weather-based index insurance for apple

Coverage level	Critical values (ton/ha)	Expected loss (ton/ha) Ave[max($y_c - y$),0]	Fair premium (rials)	Actual premium (rials)
100	29.243	9.298	32546110	36162340.44
70	20.470	3.344	11704950	13005500
50	14.621	0.890	3118460.7	3464960.33

Wakeby distribution is more suitable, and its location parameter is $\xi = 8.72$, the scale parameters are $\alpha = 18.59$ and $\beta = 0.52$, and the shape parameters are $\gamma = 0$ and $\delta = 0$.

The study reveals that ARIMA (0,1,2) is the best-fitted model for the yield forecasting, and is selected based on the minimum value of AIC and BIC. Thus, the forecasted yield of Damavand apple is about 29.243 (ton/ha) for next crop year. We use this forecasted value to calculate the critical value of yield. Here, the price of apple is 3,500 Rials that is the price determined by Agricultural Insurance Fund for apple under the current insurance scheme. Thus, we calculate the expected loss and premium by comparing the critical values of yield at three coverage levels (50, 70, and 100%) and simulated values. The results are reported in Table 9.

The expected loss at 100% coverage level is 9.298 tons/ha. Considering the total apple cultivated area in Damavand, i.e., 6,875 ha; it can be claimed that the expected loss in the whole region is 63,923.75 tons; given the important role of apple production in the country and Damavand economy, this loss is noticeable. Decreasing the coverage levels, the premium decreases as well. According to the results of Table 9, the weather-based index insurance plan premium at 100% coverage level (36,162.34044 thousand rials) is less than the current insurance premium in the crop year 2014–15 (39,170 thousand rials) and is greater than the premium in 2015–16 and 2016–17 (27,700 and 13,900 thousand rials, respectively). This diversity arises from the different essence of two insurance plans and circular and administrative style of current insurance system. In addition, in the current system, premium is considered based on the production cost, while in weather-based index insurance, the premium is computed based on the value of production or yield; furthermore, in the current apple insurance, weather factors are not considered.

4 Conclusion and Policy Recommendation

Damavand is considered as one of the biggest apple-producing region in Tehran and the whole country, which is exposed to various types of risks. The most important risk factor that results in poor crop yield is weather changes. Due to the importance of apple commercial production and its main role in the Damavand and country's horticulture economic condition, agricultural insurance was used to manage such risks. Currently, the agricultural insurance is traditionally implemented so that it has high transaction costs and is unprofitable because of the problems caused by moral hazard and adverse selection. Furthermore, regarding the decreasing amount of apple premium in the current insurance, which is 39,170 thousand rials in 2014–15 and is 27,700 and 13,900 thousand rials in 2015–16 and 2016–17, respectively, shows that the insurance system has an administrative and circular mode and does not work based on the expected loss. This issue reduces the willingness of gardeners to participate in the insurance plan.

To overcome the asymmetric information problems, and to increase the participation of gardeners in insurance, the global approach move toward application of index-based insurance, such as weather-based index insurance. In indemnity payment, weather indices are replaced instead of gardener's behavior, which prevents asymmetric information. In addition, the expected loss is as a compensation criterion that reduces transaction costs, loss assessment, bureaucracy, and scope of corruption. This is a scheme that encourages gardeners to participate in insurance plan to improve their economic and social conditions. Therefore, it is expected that the Agricultural Insurance Fund moves toward such plans to increase the economic growth of agricultural sector. In this regard, the study aims to design the weather-based index insurance as an efficient tool in Damavand apple risk management, and tries to provide an effective guide for researchers and policy-makers.

In order to achieve this goal, the information on apple yield and weather variables were collected from Iranian Agricultural Organization and Meteorological station in Damavand over the period from 1987 to 2016. In this study, according to the local experts' suggestions, we used the average temperature, cumulative rainfall index, relative humidity, and faster wind speed as weather variables. Regarding the loss factors in the region, decline in cumulative rainfall, especially in the hibernation stage, leads to a shortage of water storage in the soil and disturbs the fruit growth in the next stages. Falling temperature and cold stress are mentioned as the most important cause of damage in the region in the germination and blossoming stages; and also that according to the report of the Agricultural Insurance Fund (2015), cold and frost caused about 70% of the total loss of apple in mountainous areas, such as Damavand. In addition, the increasing relative humidity disrupts fertilization and pollination in the blossoming stage. The increasing of wind speed during fruit ripening time causes the stains on the apple. In index insurance plan, accurate calculation of expected loss is very important. Since we use various weather variables, a flexible multivariate function is applied to model the dependency structure, that is, the vine copula. Due to inadequate long time series, we use the Bayesian approach to estimate the copula parameter. The investigation results of vine copulas show that the C-vine acts better than the D-vine in explanation of the join distribution. As a result, C-vine copula is used to calculate the expected loss and premium. Thus, the expected loss is equal to 9.298 tons/ha and the corresponding premium becomes 36,162.34044 thousand rials.

Considering the experience of developing countries in implementation of weather-based index insurance, it is expected that this plan becomes successful and together with the fact that stabilizes gardeners' income, reduces their weather risks, and solves the problems of Agricultural Insurance Fund. It should be noted that this insurance may have basic risk. To solve this problem, it is better to select homogeneous regions in terms of climate and topography, and implement the pilot program in the efficient garden to calculate the expected weather loss accurately and eliminate the part of loss caused by inefficiencies of managers. In addition, it is necessary to carefully and expertly investigate the experience of developing countries that can increase the efficiency of this plan. It is advisable to implement this plan in small cases and solve its shortcomings then extend it to the whole country. Furthermore, updating the weather data and applying appropriate methods to increase the accuracy of loss measurement can encourage farmers to participate in this insurance.

Appendix

(a) Diagrams of posterior distribution of pair copulas' parameters in C-vine

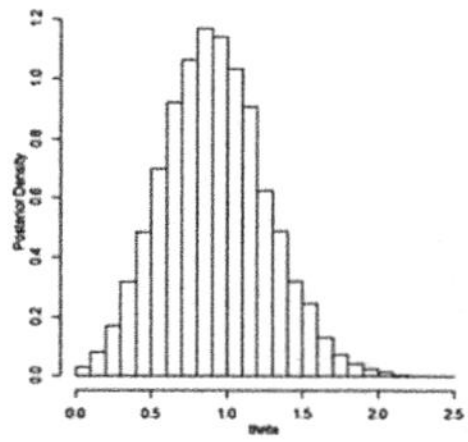

Clayton parameter ($P_{5,1}$)

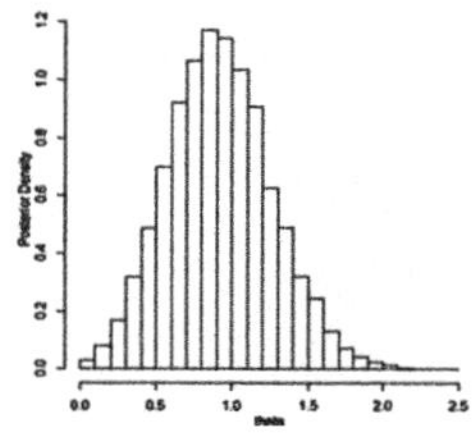

Gaussian parameter ($P_{5,2}$)

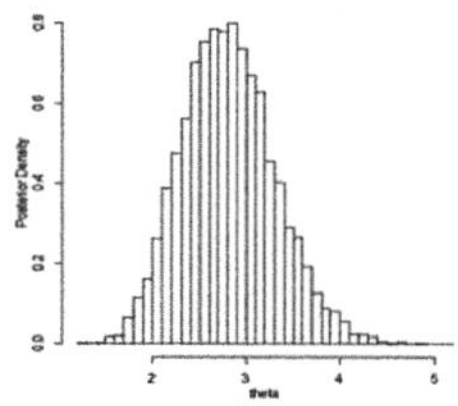

Joe -180 ° parameter ($P_{5,3}$)

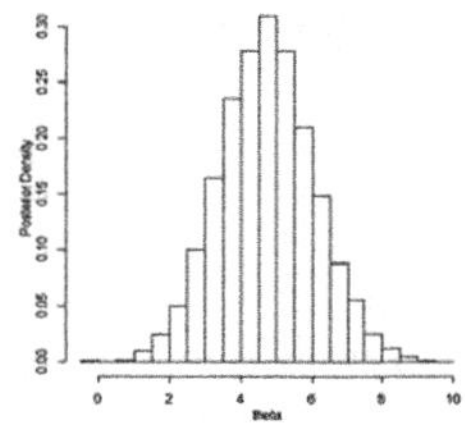

Frank parameter ($P_{5,4}$)

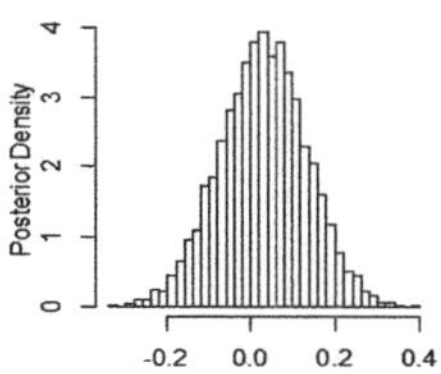

Gaussian parameter ($P_{1,2|5}$)

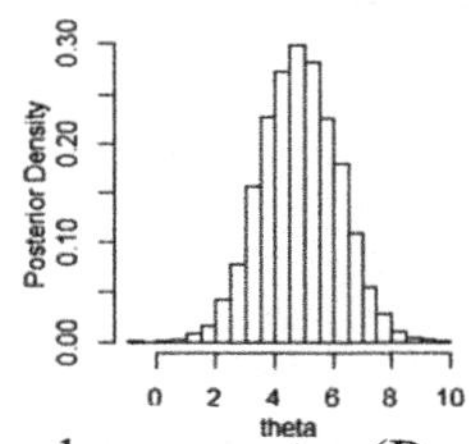

Frank parameter ($P_{1,3|5}$)

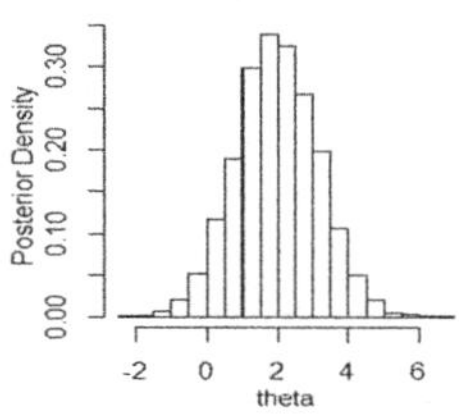

Frank parameter ($P_{1,4|5}$)

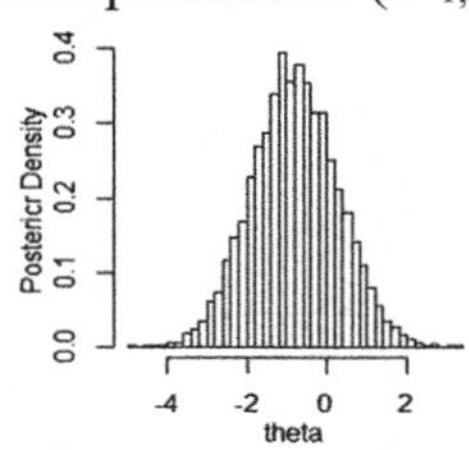

Frank parameter ($P_{2,3|5,1}$)

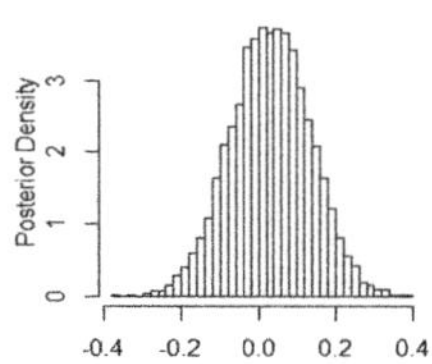

Gaussian parameter ($P_{2,4|5,1}$)

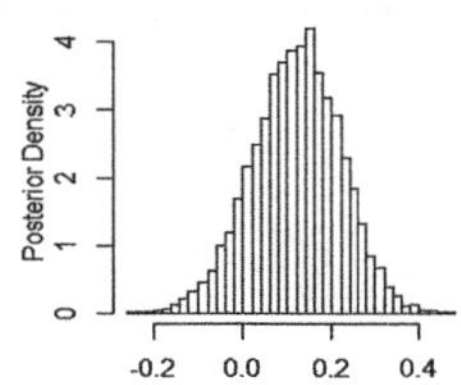

Gaussian parameter ($P_{3,4|5,1,2}$)

(b) Diagrams of posterior distribution of pair copulas parameters in D-vine

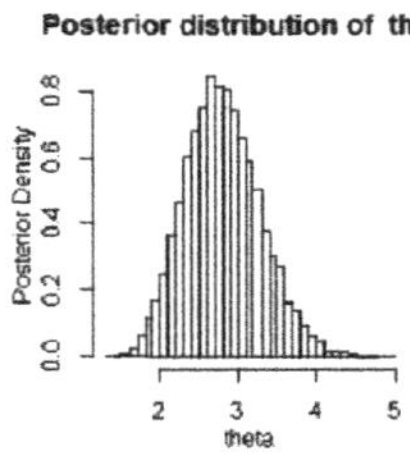

Joe180 parameter ($P_{3,5}$)

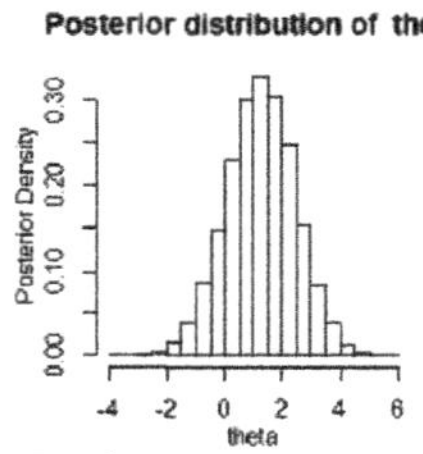

Frank parameter ($P_{4,2}$)

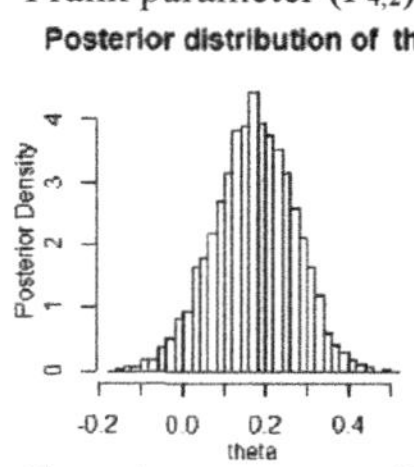

Gaussian parameter ($P_{3,4|5}$)

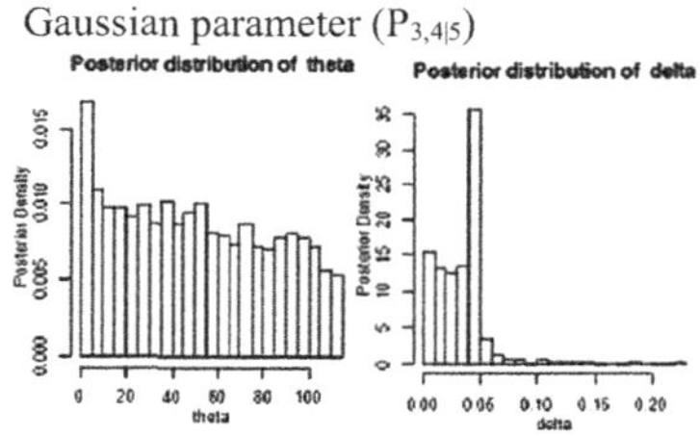

TawnII 180 parameters $P_{1,4|5,3}$
Left panel is P_1 and right panel is P_2

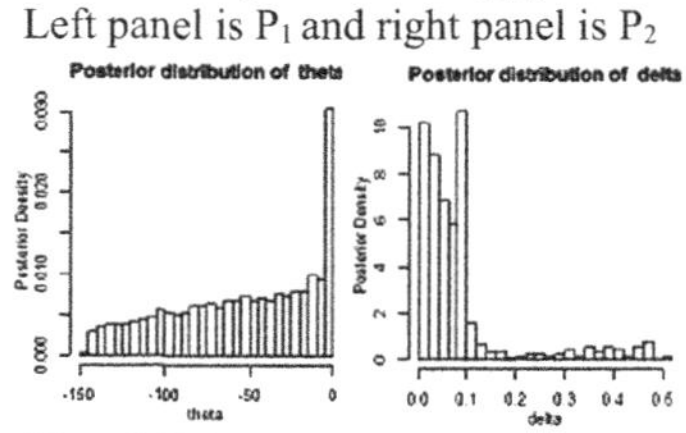

TawnI 270 parameters $P_{1,2|5,4,3}$
Left panel is P_1 and right panel is P_2

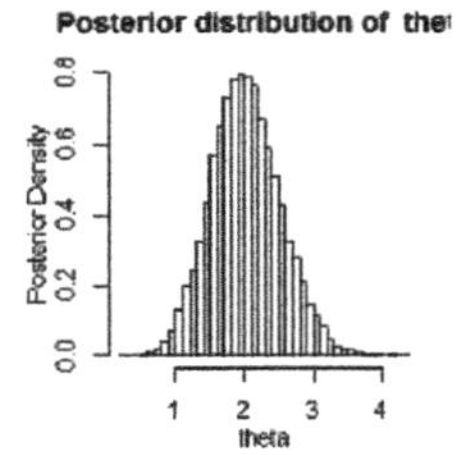

Clayton parameter ($P_{1,3}$)

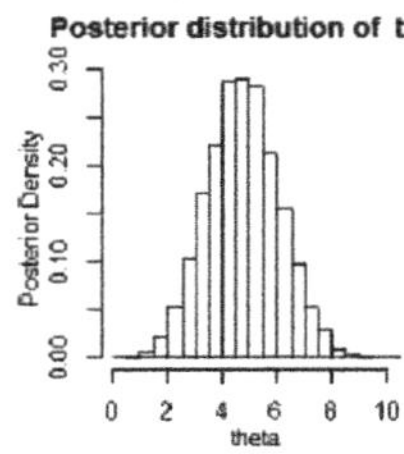

Frank parameter ($P_{5,4}$)

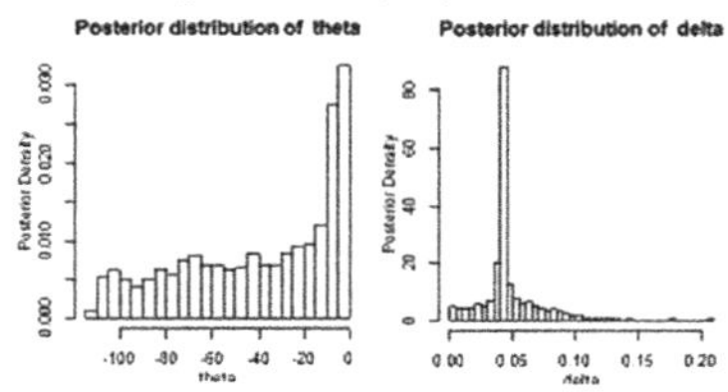

TawnI 270 parameters ($P_{1,5|3}$)
Left panel is P_1 and right panel is P_2

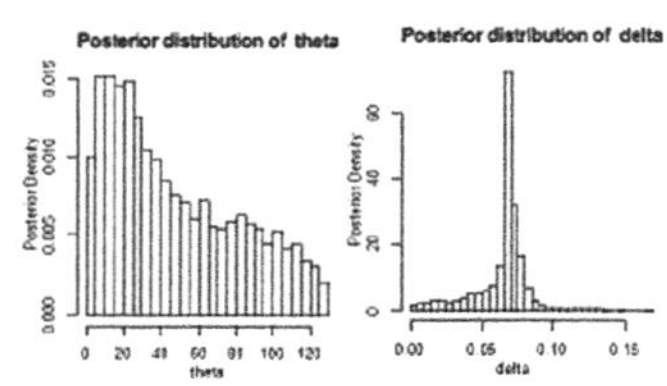

TawnII 180 parameters ($P_{5,2|4}$)
Left panel is P_1 and right panel is P_2

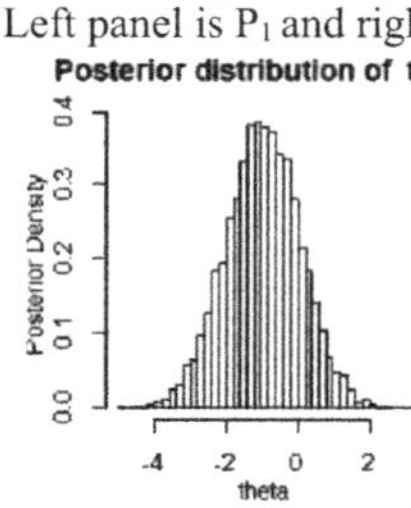

Frank parameter $P_{3,2|5,4}$

References

Aas K, Czado C, Frigessi A, Bakken H (2009) Pair-Copula constructions of multiple dependence. Insur Math Econ 44:182–198
Agricultural Insurance Fund (2015) Report on the performance of agricultural insurance fund during the recent years. Manage Plan Serv (in Persian)
Agricultural Statistics (2015) Agriculture ministry. Department of planning and economy, IT center, vol 2, 1st edn (in Persian)
Aziznasiri S (2011) Weather-based crop insurance as a viable instrument for agricultural risk management in Iran. Master of Science thesis, Allameh Tabatabai University, E.C.O. College of Insurance (in Persian)
Beatriz V, Mendes M, Eduardo F, de Melo L, Nelsen R (2005) Robust fits for copula models. UFRJ/COPPEAD
Bedford T, Cooke RM (2001) Probability density decomposition for conditionally dependent random variables modeled by vines. Ann Math Artif Intell 32:245–268
Bedford T, Cooke RM (2002) Vines: a new graphical model for dependent random variables. Ann Stat 30:1031–1068
Blanc E (2012) The impact of climate change on crop yields in Sub-Saharan Africa. Am J Clim Chan 1(1):1–13
Bokusheva R (2010) Measuring the dependence structure between yield and weather variables. ETH Zurich, Institute for Environmental Decisions
Brechmann EC, Schepsmeier U (2012) Modeling dependence with C- and D-vine copulas: the R-package C-D vine. J Stat Softw 52(3):1–27
Brechmann EC, Czado C, Aas K (2010) Truncated regular vines and their applications. Can J Stat 40(1):68–85
Burke M, Hsiang SM, Miguel E (2015) Global non-linear effect of temperature on economic production. Nature 527:235–239
Carlin BP, Louis T (2000) Bayes and empirical Bayes methods for data analysis. Chapman and Hall, New York
Chen S, Wilson WW, Larsen R, Dahl B (2013) Investing in agriculture as an asset class. Department of Agribusiness and Applied Economics Agricultural Experiment Station North Dakota State University
Conradt S, Robert F, Bokusheva R (2015) Tailored to the extremes: quantile regression for index-based insurance contract design. Agric Econ 46:1–11
Czado C, Brechmann EC, Gruber L (2014) Selection of vine copulas. Technische Universitat Munchen
Daron JD, Stainforth DA (2014) Assessing pricing assumptions for weather index insurance in a changing climate. Climt Risk Mang 1:76–91
Dell M, Jones BF, Olken BA (2014) What do we learn from the weather? The new climate-economy literature. J Econ Lit 52(3):198–204
Emmanouil KN, Nikos N (2012) Extreme value theory and mixed canonical vine copulas on modelling energy price risks. Cass Business School, City University London
Farzaneh F, Allahyari MS, Damalas CA, Seidavi A (2017) Crop insurance as a risk management tool in agriculture: the case of silk farmers in northern Iran. Land Use Pol 64:225–232 (in Persian)
Food and Agricultural Organization (FAO) (2014) FAO production year book (in Persian)
Geidosch M, Fischer M (2016) Application of vine copulas to credit portfolio risk modeling. J Risk Fin Mang 9: 4; https://doi.org/10.3390/jrfm9020004
Governor's Office of Damavand (2015) Introduction to damavand county. Available at: https://damavand.ostan-th.ir (in Persian)
Guiteras R (2007) The impact of climate change on Indian agriculture. Department of Economics, Massachusetts Institute of Technology (MIT), Mimeo
Iranian Agricultural Organization site, Tehran Province (2015) Available at: Tehran.agri-jahad.ir (in Persian)

Jie C, Li Y, Sijia L (2013) Design of wheat drought index insurance in Shandong province. Int J Hybrid Inf Technol 6(4):95–104
Joe H (1996) Families of m-variate distributions with given margins and m(m-1)/2 bivariate dependence parameters. Ins Math Stat Hayward
Kim D, Kimb JM, Liao SM, Jung YS (2013) Mixture of D-vine copulas for modeling dependence. Comput Stat Data Anal 64:1–19
Kochakzaei F, Kochakzaei A (2015) The study of weather-based index agriculture insurance in numerous different countries. International Conference on Applied Researches in Agriculture, Melard, Iran. Retrieved from http://www.civilica.com/Paper-ICARA01-ICARA01_085.html (in Persian)
Kochakzaei F, Norouzi Gh, Goudarzi M (2013) The calculation of agricultural insurance premium of rainfed wheat through precipitation index (case study: Daregaz town). Tehran, Iran. In: The 1st national conference on stable agriculture and natural resources. Proceedings of MehrArvand Higher Education Institute, Extension group of environmentalists and the Association of Iran's nature advocacy. Retrieved from http://www.civilica.com/Paper-NACONF01-NACONF01_0520.html (in Persian)
Kurowicka D, Cooke RM (2006) Uncertainty analysis with high dimensional dependence dodelling. Wiley
Leblois A, Quirion P (2010) Agricultural insurances based on meteorological indices: realizations, methods and research agenda. Downloaded from http://ideas.repec.org
Lobell DB, Field CB (2007) Global scale climate crop yield relationships and the impacts of recent warming. Env Res Lett 2(1):625–630
Lobell DB, Schlenker W, Costa-Roberts J (2011) Climate trends and global crop production since 1980. Science 333(6042):616–620
Ma J, Maystadt JF (2017) The impact of weather variations on maize yields and household income: income diversification as adaptation in rural China. Global Env Change 42:93–106
Pishbahar E, Abedi S, Dashti G, Kianirad A (2015) Weather-based crop insurance (WBCI) premium for rainfed wheat in Miyaneh county: D-Vine copula approach application. J Agric Econ 9(3):37–62 (in Persian)
Pishbahar E, Abedi S, Dashti G, KianiRad A (2019) Agricultural risk management through weather-based insurance in Iran. In: Rashidghalam M (eds) Sustainable agriculture and agribusiness in Iran. Perspectives on development in the Middle East and North Africa (MENA) region. Springer, Singapore
Robison LJ, Barry PJ (1987) The competitive firm's response to risk. Macmillan, New York
Roudier P, Sultan B, Quirion P, Berg A (2011) The impact of future climate change on West African crop yields: what does the recent literature say? Global Env Change 21(3):1073–1083
Schlenker W, Lobell DB (2010) Robust negative impacts of climate change on African agriculture. Env ResLett 5(1):123–129
Sklar A (1959) Fonctions de repartition a n dimension et leurs marges. Publications de l'Institut de Statistique de L Universite de Paris 8:299
Tao F, Zhang Z (2010) Dynamic responses of terrestrial ecosystems structure and function to climate change in China. J Geop Res: Biogeo 115(3):58–72
Welch JR, Vincent JR, Maximilian A, Moya PF, Achim D, David D (2010) Rice yields in tropical-subtropical Asia exhibit large but opposing sensitivities to minimum and maximum temperatures. Proc Nat Acad Sci USA 107(33):14562–14567
Yao J, Liu Z, Yang Q, Liu Y, Chengzhi LI, Wenfeng HU (2014) Temperature variability and its possible causes in the typical basins of the arid Central Asia in recent 130 years. Acta Geogr Sinica 69(3):291–302
Zhang Q, Zhang J, Guo E, Yan D, Sun Z (2015) The impacts of long-term and year-to-year temperature change on corn yield in China. Theor App Clim 119:77–82
Zhang T, Huang Y (2012) Impacts of climate change and inter-annual variability on cereal crops in China from 1980 to 2008. J Sci Food Agric 92(8):1643–1652

Zhao Y, Chai Z, Delgado M, Preckel P (2017) A test on adverse selection of farmers in crop insurance: results from Inner Mongolia, China. J Integr Agric 16(2):478–485

Sasan Torabi is a lecturer at Islamic Azad University of Tehran Sama and Shahr-e-rey branches. He did his B.A in Department of Economics at Allameh Tabataba'i University in Tehran and holds a M.A in economic sciences from Islamic Azad University. He is a Ph.D. student in agricultural economics at the International Campus of Ferdowsi University of Mashhad.

Arash Dourandish is an Associate Professor in Department of Agricultural Economics at Ferdowsi University of Mashhad. He holds a B.Sc. from Ferdowsi University of Mashhad and a M.Sc. and Ph.D. from University of Tehran. He spent a research course in Germany to complete his thesis at Gottingen University. His research fields include production economics, mathematical programming, agricultural risk management and insurance.

Mahmoud Daneshvar Kakhki is a Professor in the Department of Agricultural Economics at Ferdowsi University of Mashhad. He holds a B.Sc. from University of Tehran in 1976 and two M.Sc. in agricultural development (Sorbonne University, France) and in rural development (University of Montpellier, France). He completed his Ph.D. in agro-food economics at Sorbonne University in 1989. His areas of interest and research are agricultural development, agricultural policy and rural planning and development. His teaching areas are agricultural policy and development at undergraduate and postgraduate levels.

Ali Kianirad is an Assistant Professor in Agricultural Economics at Agricultural Planning, Economics and Rural Development Research Institute (APERDRI). He holds a B.Sc. in Agricultural Economics from Kerman Shahid Bahonar University and a M.Sc. and Ph.D. in Agricultural Economics from University of Tehran. He teaches Food Policy, Agricultural Policy and Natural Resource Economics at both undergraduate and post graduate levels. His areas of concentration are Agricultural Economics, Agricultural Insurance and Risk Management.

Hosein Mohammadi is an Associate Professor in the Department of Agricultural Economics at Ferdowsi University of Mashhad. He received his Ph.D. in Economics, from Allameh Tabataba'i University in Tehran. He currently teaches undergraduate and postgraduate courses in macroeconomics, international trade, international marketing and econometrics. His fields of expertise are international trade and marketing.

An Investigation on Dependency Structure Between Temperature-Humidity Index (THI) and Milk Yield

Afsaneh Nikoukar and Sasan Torabi

Abstract The heat stress and high humidity reduce volatile fatty acids in dairy cow's rumen and finally, milk production. In Iran, particularly in Tehran Province and Damavand County, as an important place in milk production, dairy cows are exposed to the risk of heat stress. Therefore, in this study, we calculated the probability and amount of milk expected loss caused by heat stress in Damavand. We collected the monthly data from Damavand Meteorological and Iranian Agricultural Organization over the period from 2012 to 2016. To increase the accuracy, the dependency structure between milk yield and Temperature-Humidity Index, which is a measurement of heat stress, is investigated using the copula function. The evaluation of different families of copula, including, Elliptical, Archimedean, and Extreme value, shows that there is a strong dependency structure between variables, so that this dependency can be represented by rotated Clayton copula better than the rest. Expected loss within a month for each dairy cow and probability of loss are 42 kg and 8.05%, respectively. Consequently, we recommend that the Agricultural Insurance Fund implement weather-based index insurance plan in warm seasons to decrease such losses so that they can stabilize cattlemen's income and protect societies' protein health.

Keywords Expected loss · Copula function · Temperature-humidity index · Dairy cow

---

A. Nikoukar (✉)
Payame Noor University, Tehran, Iran
e-mail: anikoukar57@gmail.com

S. Torabi
Department of Agricultural Economics, International Campus of Ferdowsi University of Mashhad, Mashhad, Iran
e-mail: sassan_torabi@yahoo.com

M. Rashidghalam (ed.), *The Economics of Agriculture and Natural Resources*, Perspectives on Development in the Middle East and North Africa (MENA) Region,
https://doi.org/10.1007/978-981-15-5250-2_6

1 Introduction

Milk contains essential nutrients, including, proteins, fat, vitamins, and minerals, etc., that are necessary for body growth, health, and disease prevention (Wongpom et al. 2017). Animal husbandry is one of the most important sub-sectors of agriculture and provides protein sources, i.e., milk and meat. Like other agricultural activities, animal husbandry sector is largely affected by weather conditions, so that weather factors, particularly temperature and humidity, severely affects dairy products, farmer's income stability, and their welfare (Xiu et al. 2012). One of the most critical issues of the cattlemen is heat stress and its effects on cow's milk production. The environmental factors such as high temperature, humidity, wind speed, and solar radiation cause heat stress (Dikmen and Hansen 2009; Akyuz et al. 2010). The heat stress can lead to bankruptcy. Hence, cattlemen must adopt policies to protect their products (Akyuz et al. 2010; Chen et al. 2004). Livestock insurance is one of these policies that can be an appropriate risk management tool (Valvekar et al. 2010; Xiu et al. 2012). However, since current insurance suffers from challenges such as moral hazard, adverse selection, and high transaction costs, the policy-makers in different countries are looking for new insurance plans. Weather-based index insurance is one of such successful insurance plans (Bokusheva 2010; Leblois and Quirion 2010; Jie et al. 2013; Daron and Stainforth 2014; Conradt et al. 2015)

Livestock weather-based index insurance is a low-cost insurance scheme and reduces the problems of current insurance. In this system, the contract and indemnity are based on local weather indices, like temperature, humidity, and other atmospheric factors that are available and generally reliable. Since the indemnity payment is done in the case that the intended index passes a predetermined threshold, this insurance system does not have the challenges of moral hazard and adverse selection (Bokusheva 2010; Jie et al. 2013). In livestock weather-based index insurance, it is necessary to select an appropriate index that severely affects the products. Thus, we need an index to estimate the degree of heat stress affecting dairy cows and their products. For this, a variety of indices are introduced, from which the Temperature-Humidity Index (THI) is the most efficient and common one.

This index uses dry-bulb temperature and dew point temperature (Dikmen and Hansen 2009; Akyuz et al. 2010). The THI combines the effects of relative humidity and the environmental temperature to determine the risk of heat stress for dairy cows (Gaughan et al. 2008). This index allows predicting the occurrence of heat stress probability in different conditions so that we can minimize the undesirable effects of heat stress (Dikmen and Hansen 2009; Akyuz et al. 2010).

Many scholars have used this index in their studies. For example, Peana et al. (2017) studied the effects of weather conditions on dairy cows' milk production. The results of this study showed that heat stress had decreased milk production. In Egypt, Nasr and El-Tarabany (2017) also showed that the high values of THI lead to less fat, protein, and milk yield of dairy cows, and it has a negative effect on the welfare and income of cattlemen. De Rensis et al. (2017) pointed out that the high temperature and humidity in the warm season have a negative effect on milk production and reproduction. Bickert and Mattiello (2016) studied the impact of heat stress and cold stress on the milk production of dairy cows and found that both of them can reduce the milk production and cattlemen's welfare, and even can endanger the health of livestock. In another study, Qi et al. (2015) showed that rising summer temperatures could lead to a 5–11% drop in dairy products. Jensen et al. (2014) argued that drought was the primary cause of livestock mortality in Kenya. In fact, Kenya's livestock insurance plan is an index insurance plan, which was based on the mortality index. Drought and vegetation stress correlated significantly to the mortality rate of livestock. In India, Daron and Stainforth (2014) claimed that the weather-based index insurance is an appropriate replacement for traditional insurance, and this insurance is as an aid to low-income farmers in developing countries. In similar studies like Zwaagstra et al. (2010), Akyuz et al. (2010), Bohmanova et al. (2007), and Deng et al. (2007), the researchers showed that weather indices have a significant impact on livestock production.

In Iran, the total milk production in 2015 was 9140 thousand tons, which accounted for 69.9% of total protein productions so that it had the highest share. Also, the amount of milk production for original, hybrid, and native cows were 3486, 4201, and 753 thousand tons, respectively. There are 1032.2 and 2842.15 thousand heads of native cows and calves, respectively, while the total cows and calves are about 8189.4 thousand heads. In 2015, the number of insureds in livestock sub-sector was 139,356 people that accounted for 9.5% of total insurers in the agricultural sector. The total received premium in this sub-sector was equal to 859034 million rials, and the amount of paid indemnity was 874,026 million rials (Organization S.o.I.A 2015). In this sub-sector, Agricultural Insurance Fund insures livestock against the drought, flood, hail, cold, frost, lightning, earthquake, fire, traffic, and road accidents, and livestock diseases (Keshavarzi Bank 2015).

In 2015, the Tehran Province with 966 thousand tons of milk production and 10.5% share of the total milk production was ranked third among the provinces of the country. The number of native cows and calves in this province were about 175.76 and 9.8 thousand heads, respectively. The population of insureds in livestock sub-sector in Tehran Province was 538 people, where they insured 704,466 livestock units. The number of damaged cases in livestock sub-sector was 538 units. In current livestock insurance, the premium, total received premium and paid indemnity amounts are equal to 70,375, 111,075, and 105,303 million rials, respectively. In terms of livestock production, Damavand County is one of the most important places in Tehran Province. Damavand County is located in the east of Tehran Province. According to 2011 census, the population and area of this county are about 757,500 and 188,000 ha, respectively (Governor's Office of Damavand 2015). Damavand is a mountain climate zone, so it has cold semi-arid climate in the middle part, and mountain climate in highlands part. Most of the downpour is in the form of snow. The maximum temperature in Damavand is 35 °C in summer, the minimum temperature is −14 °C in winter and the precipitation amount is 350 mm. As mentioned, this county is one of the main centers of livestock production and livestock mainly are kept in mountainous areas. In 2014, the population of light and heavy livestock in Damavand County was about 49,000 and 4,200 heads, respectively. The total amount of livestock production in Damavand was equal to 42.8 thousand tons that accounted for 3.6% of production in Tehran Province. The total amount of milk production in Damavand was 25,656 tons (Organization S.o.I.A 2015). Due to the geographical conditions of Damavand, livestock and livestock production can be severely affected by climate change and the performance can be reduced, so cattlemen demand appropriate policies of insurance. In this regard, the covered risks by livestock insurance include loss of livestock caused by drought, flood, hail, cold and frost, lightning, earthquake, fire, transportation and road accident, and various animal diseases (Keshavarzi Bank 2015).

Due to the high capacity of weather-based index insurance, governments and international organizations have widely considered and supported it. In Iran, Pishbahar et al. (2019) studied the weather-based index insurance for rain-fed wheat during 1987–2013, and showed that this insurance system can solve the traditional insurance problem, i.e., asymmetric information. Consequently, they pointed that such insurance are more useful and efficient for agricultural products in Iran. Therefore, it is necessary to apply this system in other developing countries like Iran. In Iran, since one of the most important damaging factor in animal husbandry activities is because of weather risky events, and additionally traditional agricultural insurance plans have problems such as unprofitability of the Agricultural Insurance Fund, high transaction costs, and the challenges of adverse selection and moral hazard; it is necessary to design weather-based index insurance for this sub-sector. To provide an appropriate insurance pattern, we must investigate the dependency structure between milk yield and damaging factor. Hence, we try to use a proper method for this issue. In this study, since Damavand has a unique role in milk production and has high milk loss caused by the risk of heat stress particularly in summer, we calculate the probability and amount of milk expected loss caused by heat stress in Damavand.

The rest of this paper is organized as follows. Section 2 discusses the methodology and is followed by the empirical results and discussion in Sect. 3. Section 4 gives a conclusion.

2 Methodology

Heat stress has harmful effects on the health and milk production of dairy cows so that it reduces feeding, and thus reduces the production of volatile fatty acids in the rumen and significantly reduces milk production. As mentioned before, we can measure heat stress using the temperature-humidity index. This index is calculated as follows:

$$\mathrm{THI} = T_{\mathrm{db}} + (0.36 \times T_{\mathrm{dp}}) + 41.2 \tag{1}$$

where T_{db} is the dry-bulb temperature, and T_{dp} is the dew point temperature that contains humidity information. When THI is less than 72, it indicates the absence of heat stress, while it exceeds the threshold value, heat stress happens, and milk production decreases (Akyuz et al. 2010).

Although heat stress is inevitable, the prediction of its occurrence probability and corresponding expected loss can provide appropriate information for a politician to adopt suitable risk management solutions. To determine the probability of occurrence and the expected loss caused by heat stress, it is necessary to measure the dependency structure between milk yield and THI; therefore, it has special importance to measure the dependency structure accurately to simulate and calculate the loss so that it will be coincident with actual data. Nowadays, the copula functions are widely used in this field. These functions were first introduced by Sklar (1959), and now they are used extensively in statistics, economics, finance, and insurance to measure the dependency and risk. A copula function, $C : [0, 1]^d \rightarrow [0, 1]$, connects the marginal distributions together to form a joint distribution. In other words, any joint distribution consists of two parts: the behavior of individual variables or the marginal distributions and the dependency structure of them. In fact, this dependency structure is explained by a copula. A copula, C, is a multivariate distribution with standard uniform marginal distributions, $U(0,1)$, on [0,1]. Thus, according to Sklar's theorem, a joint distribution function, F, with marginal distributions, $F_i(x_i) = u_i$, is defined as

$$F(x_1, \ldots, x_n) = C(F_1(x_1), \ldots, F_n(x_n)) \tag{2}$$

Inverting the marginal distribution functions, $F_i^{-1}(u_i)$, the copula function is given by

$$C(u_1, \ldots, u_n) = F(F_1^{-1}(u_1), \ldots, F_n^{-1}(u_n)) \tag{3}$$

If the joint distribution and marginal distributions are continuous and strictly increasing, then the density function of joint distribution and copula can be obtained by the chain rule as

$$f(x_1, \ldots, x_n) = c_{1\ldots n}(F_1(x_1), \ldots, F_n(x_n)) \prod_{i=1}^{n} f_i(x_i) \tag{4}$$

$$c(u_1, \ldots, u_n) = \frac{f(F_1^{-1}(u_1), \ldots, F_n^{-1}(u_n))}{\prod_{i=1}^{n} f_i(F_i^{-1}(u_i))} \tag{5}$$

where f_i is the marginal density function, f is the joint density function, and c is the copula density function (Aas et al. 2009).

In general, the copulas are divided into implicit and non-implicit. Implicit copulas such as Gaussian and Student's t are directly derived from the well-known multivariate distribution functions and are known as Elliptical (Fischer 2003). In contrast to them, the non-implicit copulas that are divided into Archimedean and Extreme value are generated by the generator and the dependence function, respectively (Mendes et al. 2007). The generator function, $\phi : [0, 1] \to [0, \infty)$, is a continuous strictly decreasing convex function such that $\phi(1) = 0$ and $\phi^{[-1]}$ is the pseudo-inverse of a generator function. The Archimedean copulas can be represented by the definition of a proper generator function as

$$c(u_1, \ldots, u_n)) = \phi^{-1}\left(\sum_{i=1}^{n} \phi(u_i)\right) \tag{6}$$

The most famous Archimedean copulas are Frank with $\phi(u) = -\ln[(e^{-\theta u} - 1)/(e^{-\theta} - 1)]$, Clayton with $\phi(u) = (u^{-\theta} - 1)/\theta$, Gumbel with $\phi(u) = -\ln(u)^{\theta}$, and Joe with $\phi(u) = -\ln(1 - (1 - u)^{\theta})$. θ is the dependency parameter and has different domains in different types of the copula (Fischer 2003). Similarly, the dependence function, A:[0,1] $\to$ [1/2, 1], is a convex function such that $\max(u, 1 - u) < A(u) < 1$ for all $u \in [0, 1]$. The Extreme value copulas have a complex form and are max-stable, i.e., if $(X_1,Y_1), (X_2,Y_2), \ldots,(X_n,Y_n)$ are identical independent distribution (i.i.d) random pairs from an Extreme value copula and $M_n = \max\{X_1, \ldots,X_n\}$, $N_n = \max\{Y_1,\ldots,Y_n\}$, there is a copula associated with the random pair (M_n, N_n). A 2-dimension Extreme value copula can be written as

$$C(u_1, u_2) = \exp\left\{\log(u_1 u_2) A\left(\frac{\log(u_1)}{\log(u_1 u_2)}\right)\right\} \tag{7}$$

The most commonly used families of these copulas are the Gambel with $A(u) = (u^{\delta} + (1 - u)^{\delta})^{1/\delta}$, Galambos with $A(u) = 1 - (u^{-\delta} + (1 - u)^{-\delta})^{-1/\delta}$, Husler and Reiss with $A(u) = u\Phi\left[\delta^{-1} + \frac{1}{2}\delta \log\left(\frac{u}{1-u}\right)\right] + (1 - u)\Phi\left[\delta^{-u} - \frac{1}{2}\delta \log\left(\frac{u}{1-u}\right)\right]$, and Tawn with $A(u) = [\delta^{\theta}(1 - u)^{\theta} + \rho^{\theta} u^{\theta}]^{\frac{1}{\theta}} + (\delta - \rho)u + 1 - \delta$. In the Husler and Reiss copula, Φ denotes the CDF of standard normal distribution, and δ is the dependency parameter in these copulas (Mendes et al. 2007). Since some of these copulas do not allow negative dependence, the rotated versions of them have been introduced. In fact, we can rotate them by 180°, 90°, and 270° to examine other types of dependencies (Brechmann and Schepsmeier 2012). Selection of the most suitable copula family for our dependency structure is based on the minimized value of AIC and BIC (Bokusheva 2014).

2.1 *Estimation of the Copula Parameter*

The Genest and Rivest approach is one of the common tools in copula parameter estimation. This approach does not need to determine the marginal distribution (Schulte-Geers and Berg 2011). To use this approach, first, we select a random sample from a pair of milk yield (Y) and THI. The measure of concordance between two variables is represented by Π_c as follows:

$$\Pi_c = P((Y_i - Y_j)(\text{THI}_i - \text{THI}_j) > 0), \quad i \neq j \tag{8}$$

where P represents the probability. In fact, Π_c measures the shift of Y and THI in one direction, and $\Pi_c = 1$ shows a complete concordance. In other words, if the THI increases, Y will increase as well. Similarly, the measure of discordance between the two variables is shown by Π_d as follows:

$$\Pi_d = P((Y_i - Y_j)(\text{THI}_i - \text{THI}_j) < 0), \quad i \neq j \tag{9}$$

This criterion measures the shift of two variables in the opposite direction. $\Pi_d = 1$ represents a complete discordance between two variables, and if the THI increases, Y will surely decrease. Π_d is equal to $1 - \Pi_c$. A common approach that measures the amount of concordance and discordance between two random variables is the correlation coefficient of the Kendall's Tau and can be obtained as

$$\tau = \Pi_c - \Pi_d = 2\Pi_c - 1 \tag{10}$$

This coefficient is bounded to the interval [−1,1]. $\tau = 0$ shows the independence of two variables (Behboudiyan 2008). Genest and Rivest (1993) showed that there is a relationship between the Kendall's Tau and the copula parameter as

$$\tau = 4\int_0^1\int_0^1 C(u_1, u_2), c(u_1, u_2)du_1du_2 - 1 \tag{11}$$

Therefore, by Kendall's Tau, we can determine the copula parameter (Goodwin et al. 2011).

2.2 *Tail Dependency*

To determine the dependency structure, it is necessary to estimate the tail dependency of two variables by the estimated copula parameter. Tail dependency measures the correlation of variables at the tails of the distribution. This criterion has two dependency coefficients that are known as upper and lower tail dependencies, and are expressed as

$$\lambda_U = \lim_{q \to 1} \frac{1 - 2q + C(q,q)}{1 - q}$$
$$\lambda_L = \lim_{q \to 0} \frac{C(q,q)}{q} \tag{12}$$

If $\lambda_U \in (0, 1]$, C has an upper tail dependency; and if $\lambda_U = 0$, no upper tail dependency exists. Similarly, if $\lambda_L \in (0, 1]$, C has a lower tail dependency; and there is no lower tail dependency if $\lambda_L = 0$. Different types of copulas have various tail dependencies that are dependent on the relationship between the copula parameter and tail dependency (Fischer 2003).

2.3 Determination of Expected Loss

To determine the dependency structure, various types of copula are investigated, and the most appropriate form to explain the joint distribution of milk yield and THI is selected. Then, we generate simulated data from the joint distribution. To calculate the expected loss, we compare the simulated milk yield and its predicted value. The Monte Carlo simulation method is used to simulate data from the copula function. In this method, using a bivariate copula function for milk yield and THI, we generate pairs (y, thi) of observations of [0,1] uniformly distributed r.v.s Y and THI, where its joint distribution is C. For this purpose, the conditional distribution function is defined as $C_{\text{thi}}(y) = P(Y \leq y|\text{THI} = \text{thi})$. Then, we generate the desired pair (y, thi) in the following way:

- Generate two independent uniforms (thi, w) $\in [0, 1]$ to draw thi.
- Compute the inverse function of $C_{\text{thi}}(y)$.
- Set $y = C_{\text{thi}}^{-1}(w)$ to draw y (Nelsen 2007).

It should be noted that the required data has approximately uniform marginal on [0,1] known as copula data. In general, the data do not have uniform margins. Therefore, we can transform them using the empirical cumulative distribution function. The simulated data have uniform margins as well; hence, to convert them to standard form, we use the inverse of the theoretical cumulative distribution that has the most conformity with milk yield distribution. As mentioned before, to calculate the expected loss, the simulated yield must be compared with its predicted value of ARIMA process. Loss occurs when the predicted value of milk yield (y_f) is higher than the simulated values (y_i). Therefore, the expected loss amount is equal to their deviations average (Pishbahar et al. 2015).

$$L = \text{Ave}[\max(y_f - y_i, 0)] \tag{13}$$

3 Empirical Results and Discussion

In this study, the required weather variables including dry-bulb temperatures and dew point temperature (°C) were collected monthly over the period from 2012 to 2016. Holstein milk yield (kg) was also considered for the same time in Damavand County.

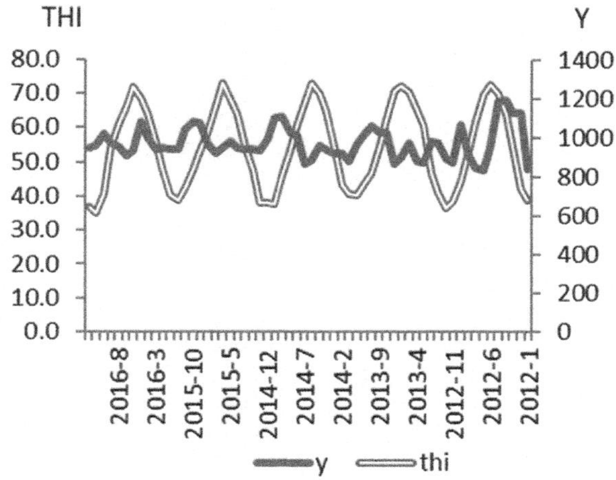

Fig. 1 milk yield and THI during 2012–16

Weather variables and milk yield were collected from Damavand Meteorological Organization and Iranian Agricultural Organization, respectively.

To evaluate the behavior of two variables, the milk yield and the THI time series are presented in Fig. 1. According to Fig. 1, the THI has a sinusoidal behavior. In other words, this variable shows a fluctuating behavior in different seasons of the year so that the milk yield has a sinusoidal behavior as well. The important issue is that the maximum points of the THI series correspond to the minimum points of the milk yield.

The descriptive results of data are reported in Tables 1 and 2. The average milk yield in Damavand is equal to 970.5 kg. The maximum amount of milk production associated with April is 1192.18 kg in 2012, while the minimum amount is about 830.4 kg that corresponds to July. The averages of THI in spring, summer, autumn, and winter are equal to 60.83, 68.78, 47.07, and 40.79, respectively. In addition, the average of milk yield that corresponds to these THI is equal to 981.99, 945.91, 1003.24, and 950.82 kg, respectively. According to Fig. 1, it is clear that as winter passes, the THI and milk yield increase. In winter, due to cold stress, a significant amount of energy is used to produce heat, and thus the milk production decreases. In May, the increasing THI increases the milk yield, but when the THI reaches its critical level in July, milk production reaches its minimum level; therefore, the average of THI in this month is about 72.59 and milk yield average is equal to 875.76 kg. Consequently, the results indicate that in summer, the heat stress is an important factor in milk yield loss.

First, we measure the Kendall's Tau to examine the independence of two variables. The value of this coefficient between the milk yield and the THI is −0.59, which indicates a strong negative relationship. Therefore, the two variables are not independent. Next, we investigate different types of Elliptical, Archimedean, and Extreme value copulas to examine the dependency structure, and to estimate the joint distribution function between the milk yield and the THI. We decide to use the Gaussian copula

Table 1 The average monthly THI

Year	Month												Annual average
	Jan	Feb	March	April	May	June	July	Aug	Sept	Oct	Nov	Dec	
2012	37	34.9	40.8	54.2	61	66.1	72.1	68.6	64	55.9	47.8	40.4	53.57
2013	38.7	42.4	47.4	53.1	58.6	66.2	73.1	68.1	63.9	54.5	47.4	37.9	54.27
2014	38	37.6	45.9	53.4	60.7	67.2	72.9	69.9	64.8	54.4	43.5	40.8	54.09
2015	40.3	43.2	46.6	54.4	62.4	70.7	72.3	70.5	65.1	60.7	46.9	40.9	56.18
2016	36.6	38.3	44	53.4	61.6	69.5	72.5	70.1	63.8	54	42.1	39	53.75
Average	38.12	39.28	44.94	53.7	60.86	67.94	72.59	69.44	64.32	55.9	45.54	39.8	54.372
Max	40.3	43.2	47.4	54.4	62.4	70.7	73.1	70.5	65.1	60.7	47.8	40.9	56.18
Min	36.6	34.9	40.8	53.1	58.6	66.1	72.1	68.1	63.8	54	42.1	37.9	53.57

Table 2 The average monthly milk production per each dairy cow (kg)

Year	Month												Annual average
	Jan	Feb	March	April	May	June	July	Aug	Sept	Oct	Nov	Dec	
2012	839.4	1126.8	1126.8	1192.2	1186.2	948.9	830.4	848.1	907.5	1067.4	871.8	889.8	986.4
2013	970.2	975.6	873.3	876.9	972.9	897.3	862.5	1024.2	1024.2	1059	1014.3	964.8	959.7
2014	877.2	921.6	916.5	936.3	957.6	882.6	862.5	1014.3	1023.6	1106.4	1098.6	990.6	965.7
2015	931.2	938.1	936	945.6	981	942	917.1	968.7	1073.1	1080.6	1036.5	940.5	947.4
2016	939.3	947.7	942	998.7	1081.2	929.1	906.3	951.6	974.4	1019.7	960.6	947.1	966.6
Average	911.46	981.96	958.92	989.94	1035.78	919.98	875.76	961.38	1000.56	1066.62	996.36	946.6	970.56
Max	970.2	1126.8	1126.8	1192.2	1186.2	948.9	917.1	1024.2	1073.1	1106.4	1098.6	990.6	986.4
Min	839.4	921.6	873.3	876.9	957.6	882.6	830.4	848.1	907.5	1019.7	871.8	889.8	959.7

Table 3 Copula parameters and tail dependency estimation

Copula family	Parameter	Tail dependency (upper, lower)
Gaussian	−0.09	(0,0)
Frank	−0.53	(0,0)
Clayton 90&270	−0.12	(0,0)
Gumbel 90&270	−1.06	(0.08,0)
Joe 90&270	−1.11	(0.13,0)
Galambos	0.27	(0.08,0)
Husler-Reiss	0.56	(0.07,0)
Tawn	0.17	(0.09,0)

because the degree of freedom for Student's t copula is greater than 30. The negative values of Kendall's Tau prevent the use of some Archimedeans' families, including, Clayton, Gombel, Joe, and their rotated versions by 180°. Therefore, as mentioned before, the rotated versions of them by 90° and 270° can be used. The estimation results of copula parameters and their tail dependency are presented in Table 3.

Since the domain of copula parameters is different, their estimated amount cannot be interpreted, but they show that there is a relationship between the milk yield and the THI. The three Gaussian, Frank, and negative rotated Clayton copulas do not have a tail dependency. The rest exhibit a strong dependency in the upper tail. In other words, given the upper tail dependency, it can be claimed that the milk yield responds more to the large amount of the THI than its mean and small amounts. Hence, heat stress affects more negatively than cold stress. The contour plots for each type of copula are presented in Fig. 2, which shows the correlation of variable.

The presented contour plots affirm the results of Table 3 so that they are symmetric for Gaussian, Frank, and negative rotated Clayton copulas, and are asymmetric for the rest. Most of the lines are seen at the upper tails; it is an evident of high correlation at the upper tail there. This issue shows that the heat stress effect is more than other situations. AIC and BIC are used to select the appropriate joint distribution function between the milk yield and the THI, and the results are reported in Table 4. The negative rotated Clayton copula has the minimum value of AIC and BIC, so it can perform better than others to simulate and determine the expected loss.

Thus, regarding the selected copula for the variables as the joint distribution function, we simulate random observations for the milk yield that is conditioned on the THI. The simulated data are in [0,1]. Therefore, the copula data are converted by the appropriate theoretical distribution function which is selected by the Kolmogorov–Smirnov, Anderson–Darling, and Chi-Squared tests. The result for distribution selection is presented in Table 5. The statistic of these tests indicates that the generalized logistics distribution ($y \sim \text{gen.}\log(k = 0.15, \sigma = 0.04, \mu = 0.95)$) with shape parameter of $k = 0.15$, scale parameter of $\sigma = 0.04$, and location parameter of $\mu = 0.95$ is more coincidence with the distribution of the milk yield. Therefore, this distribution is used to convert the milk yield to its standard form. The average value of the simulated milk yield is 947 kg.

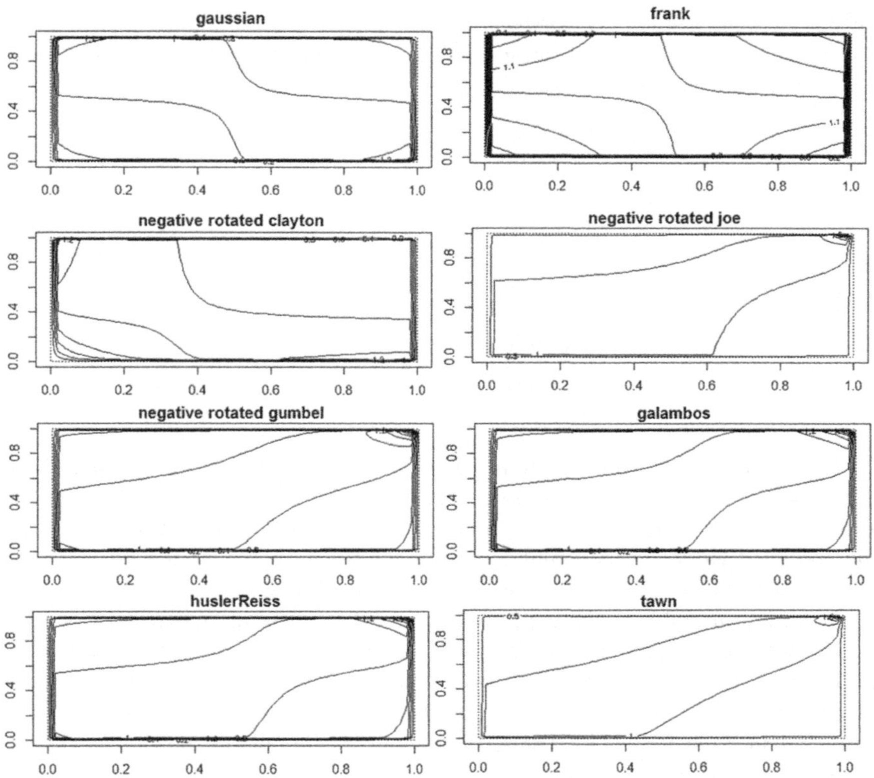

Fig. 2 The copula's contour plat

Table 4 Copula selection

Copula family	AIC	BIC
Gaussian	−11.78	−6.04
Frank	−9.32	−5.49
Clayton 90&270	−6.21	−2.04
Gumbel 90&270	−45.66	−33.84
Joe 90&270	−52.68	−40.87
Galambos	−51.24	−39.43
Husler-Reiss	−53.63	−41.81
Tawn	−41.17	−29.35

Table 5 Theoretical distribution selection

Selected distribution		Kolmogorov–Smirnov	Anderson–Darling	Chi-squared
Gen-logistic	Statistics	0.06	0.29	4.12
	p-value	0.96	–	0.53
	Critical value (α = 5%)	0.17	2.51	11.07

Table 6 Expected loss and its probability

Critical value (kg)	967
Expected loss (kg)	42
Loss probability (%)	8.05

The prediction of the milk yield is made by ARIMA (0,1,1) process, which shows that yield (Y) is stationary at first difference. According to this process, the predicted value of the milk yield is 967 kg per dairy cow in Damavand County. Thus, comparing the predicted yield and the simulated observations, we calculate the expected loss and the probability of loss. The results are shown in Table 6.

According to Table 6, the expected loss per dairy cow in Damavand County is 42 kg. In the case that the THI reaches its critical value and causes heat stress, the probability of loss will be 8.05%. Besides, since the total number of dairy cows in Damavand is equal to 2207 heads, it can be expected that the loss caused by heat stress will be about 92,694 kg. Due to the specific role of milk in nutrition and health, this expected loss is very high.

4 Conclusion and Policy Recommendations

The gradual warming and weather change cause stress for dairy cows so that the heat besides the humidity increases the stress. This stress causes respiratory infections, reduces feeding, volatile fatty acid in the rumen, and finally milk production. Although Damavand County is in the mountainous climate, in summer the temperature-humidity index exceeds its critical value and causes heat stress in dairy cows, and reduces milk production. Therefore, due to the importance of this issue, the expected loss caused by heat stress in Damavand dairy cows was evaluated. Data were collected monthly from Damavand Meteorological Organization and Iranian Agricultural Organization between 2012 and 2016. The dependency structure between THI and the milk yield was investigated using different types of Elliptical, Archimedean, and Extreme value copulas.

The result of tail dependency showed that the milk yield is more affected by the large and critical values of THI. In other words, heat stress was more stressful than the cold stress. The negative rotated Clayton copula can explain the joint distribution function better than the rest. This copula shows a strong negative correlation between the milk yield and the THI. Therefore, as the THI passes its critical value, the probability of heat stress occurrence increases, and the milk yield decreases. These results are consistent with previous findings by Linvill and Pardue (1992), West et al. (2003), Deng et al. (2007), Akyuz et al. (2010) and Taghavi et al. (2014). Regarding the negative rotated Clayton copula, the expected loss of milk yield for heat stress is 42 kg per dairy cow. Furthermore, the probability of loss caused by heat stress is about 8.05% that is higher in warm months.

To reduce the harmful effects of heat stress, it is necessary to predict the occurrence of heat stress and inform the cattlemen. The government can also give cattlemen appropriate facilities to buy mechanical fans and other proper weather ventilation equipment during the warm seasons. In developed countries, they apply weather-based index insurance to stabilize the income of cattlemen and increase the milk production. Given the role of milk in societies' health, it is expected that Agricultural Insurance Fund adopt this risk management tool as well. The proper and efficient implementation of this insurance system depends on the correct calculation of the expected loss. Since, milk is produced in heterogeneous climatic and environmental areas, homogeneous areas should be included in this plan. In addition, to reduce the error rate, new methods of dependency structure measurement should be applied and the weather data should be collected at the farm level. Implementation of this insurance scheme has high initial costs to install pilots and measure weather variables, but leads to long-term benefits in reducing the risk of heat stress and increasing milk production. Finally, it is necessary to carry out an expert examination of the experiences of other countries to reach successful results and to use the best and least challenging methods used by others that are compatible with the conditions of each region in Iran.

References

Aas K, Czado C, Frigessi A, Bakken H (2009) Pair-copula constructions of multiple dependence. Insur Math Econ 44:182–198

Akyuz A, Boyaci S, Cayli A (2010) Determination of critical period for dairy cows using temperature humidity index. J Anim Vet Adv 9:1824–1827

Behboudiyan J (2008) Non-parametric statistics. Fifth Edition, Shiraz University (in Persian)

Bickert WG, Mattiello S (2016) Stress in dairy animals: cold stress: management considerations. Ref Mod F Sci 2:555–560

Bohmanova J, Misztal I, Cole J (2007) Temperature-humidity indices as indicators of milk production losses due to heat stress. J Dairy Sci 90:1947–1956

Bokusheva R (2010) Measuring the dependence structure between yield and weather variables. ETH Zurich, Institute for Environmental Decisions

Bokusheva R (2014) Improving the effectiveness of weather-based Insurance: an application of copula approach. ETH Zurich, Institute for Environmental Decisions

Brechmann E, Schepsmeier U (2012) Modeling dependence with C-and D-vine copulas. The R-package CDVine. Available online: http://cran.r-project.org/web/packages/CDVine/vignettes/CDVine-package.pdf. Accessed on 1 Sept 2012

Chen G, Roberts MC, Thraen C (2004) Weather derivatives in the presence of index and geographical basis risk: hedging dairy profit risk. In: NCR-134 conference on applied commodity price analysis, forecasting, and market risk management, St. Louis, MO

Conradt S, Finger R, Bokusheva R (2015) Tailored to the extremes: quantile regression for index-based insurance contract design. Agric Econ 46:537–547

Daron JD, Stainforth DA (2014) Assessing pricing assumptions for weather index insurance in a changing climate. Clim Risk Manage 1:76–91

De Rensis F, Lopez-Gatius F, García-Ispierto I, Morini G, Scaramuzzi R (2017) Causes of declining fertility in dairy cows during the warm season. Theriogenology 91:145–153

Deng X, Barnett BJ, Vedenov DV, West JW (2007) Hedging dairy production losses using weather-based index insurance. Agric Econ 36:271–280

Dikmen S, Hansen P (2009) Is the temperature-humidity index the best indicator of heat stress in lactating dairy cows in a subtropical environment? J Dairy Sci 92:109–116

Fischer MJ (2003) Tailoring copula-based multivariate generalized hyperbolic secant distributions to financial return data: An empirical investigation. Diskussionspapiere//Friedrich-Alexander-Universität Erlangen-Nürnberg

Gaughan J, Mader TL, Holt S, Lisle A (2008) A new heat load index for feedlot cattle. J Anim Sci 86:226–234

Genest C, Rivest L-P (1993) Statistical inference procedures for bivariate Archimedean copulas. J Am Stat Assoc 88:1034–1043

Goodwin BK, Holt MT, Onel G, Prestemon JP (2011) Copula-based nonlinear models of spatial market linkages. Agric. and Applied Econ. Association, Pittsburgh, PA, pp 24–26

Governor's Office of Damavand (2015) Introduction to Damavand County. Available at: https://damavand.ostan-th.ir (in Persian)

Jensen ND, Barrett CB, Mude AG (2014) Basis risk and the welfare gains from index insurance: evidence from northern Kenya

Jie C, Li Y, Sijia L (2013) Design of wheat drought index insurance in Shandong Province. Int J Hybrid Inf Technol 6:95–104

Keshavarzi Bank (2015) Tehran province headquarter (in Persian)

Leblois A, Quirion P (2010) Agricultural insurances based on meteorological indices: realizations, methods and research agenda

Linvill D, Pardue F (1992) Heat stress and milk production in the South Carolina Coastal Plains 1. J Dairy Sci 75:2598–2604

Mendes BV, De Melo EF, Nelsen RB (2007) Robust fits for copula models. Commun Stat Simul Comput 36:997–1017

Nasr MA, El-Tarabany MS (2017) Impact of three THI levels on somatic cell count, milk yield and composition of multiparous Holstein cows in a subtropical region. J Therm Biol 64:73–77

Nelsen RB (2007) An introduction to copulas. Springer Science & Business Media

Organization S.o.I.A (2015) Statistic of Iranian Agricultural Organization. Agriculture Ministry. IT center, Department of planning and economy (in Persian)

Peana I, Francesconi AHD, Dimauro C, Cannas A, Sitzia M (2017) Effect of winter and spring meteorological conditions on milk production of grazing dairy sheep in the Mediterranean environment. Small Rum Res 153:194–208

Pishbahar E, Abedi S, Dashti Gh, Kianirad A (2015) Investigating weather-based insurance index of the rain-fed wheat: D-vine copula approach. J Agric Econ 9:37–62 (in Persian)

Pishbahar E, Abedi S, Dashti G, KianiRad A (2019) Agricultural risk management through weather-based insurance in Iran. In: Rashidghalam M (eds) Sustainable agriculture and agribusiness in Iran. Perspectives on development in the Middle East and North Africa (MENA) Region. Springer, Singapore

Qi L, Bravo-Ureta B, Cabrera V (2015) From cold to hot: climatic effects and productivity in Wisconsin dairy farms. J Dairy Sci 98:8664–8677

Schulte-Geers M, Berg E (2011) Modelling farm production risk with copulae instead of correlations. In: 2011 International Congress, August 30–September 2, 2011, Zurich, Switzerland. European Association of Agricultural Economists

Sklar M (1959) Fonctions de repartition an dimensions et leurs marges. Publ Inst Stat Univ Paris 8:229–231

Taghavi H, Naserian A, Valizadeh R (2014) Determination of critical weather periods in management of dairy cows in Northeast of Iran using temperature humidity index. Iran Quart Anim Sci 6:295–303 (in Persian)

Valvekar M, Cabrera V, Gould B (2010) Identifying cost-minimizing strategies for guaranteeing target dairy income over feed cost via use of the Livestock Gross Margin dairy insurance program. J Dairy Sci 93:3350–3357

West J, Mullinix B, Bernard J (2003) Effects of hot, humid weather on milk temperature, dry matter intake, and milk yield of lactating dairy cows. J Dairy Sci 86:232–242

Wongpom B, Koonawootrittriron S, Elzo MA, Suwanasopee T (2017) Milk yield, fat yield and fat percentage associations in a Thai multibreed dairy population. Agric Nat Res 51:218–222

Xiu F, Xiu F, Bauer S (2012) Farmers' willingness to pay for cow insurance in Shaanxi province, China. Procedia Econ Fin 1:431–440

Zwaagstra L, Sharif Z, Wambile A, Leeuw JD, Said M, Johnson N, Njuki J, Ericksen P, Herrero M (2010) An assessment of the response to the 2008–2009 drought in Kenya

Afsaneh Nikoukar is an Associate Professor in Agriculture Department of Payam-e Noor University. She holds a B.Sc. from Ferdowsi University of Mashhad and a M.Sc. and Ph.D. from University of Tehran. She spent five months in Germany as a Research Scholar to complete her thesis at Gottingen University. Her areas of expertise include Agricultural Policies, Agricultural Markets, Agricultural Risk Management and Insurance. Also she teaches and supervises M.Sc. and Ph.D. Students in Environmental Economics and Natural Resource Economics topics.

Sasan Torabi is a lecturer at Islamic Azad University of Tehran Sama and Shahr-e-rey branches. He did his B.A in Department of Economics at Allameh Tabataba'i University in Tehran and holds a M.A in economic sciences from Islamic Azad University. He is a Ph.D. student in agricultural economics at the International Campus of Ferdowsi University of Mashhad.

Adoption of IPM by Farmland Owners and Non-owners: Application of Endogenous Switching Copula Approach

Sahar Abedi, Pariya Bagheri, and Esmaeil Pishbahar

Abstract Khuzestan Province is one of the largest hubs of agricultural production, and consequently is one of the largest consumers of chemical pesticides and fertilizers in the country. Integrated pest management can be an effective step toward reducing pesticide use, protect human health, and the environment. Due to the fact that owner and non-owner have different economical–social conditions, it is expected that they have two different sets of priorities in implementing of the operation. To study the effective factors on willingness to pay to reduce risks of environmental pesticides for two groups, endogenous switching method leads to better results. The problem with this method is incorrect normal distribution assumption for residuals. Therefore, in this study to cope with this problem, we applied endogenous switching copula approach which allows us to use different marginal distributions and leads to accurate results. The results showed that the logistic distribution for decision equation's residual and Student's *t*-distribution for willingness to pay equation's residual are better than normal distribution. In addition, the average treatment effect results showed that owners have more willingness to pay than non-owners; hence, different factors effect on willingness to pay for two groups. The knowledge factor has a positive effect on the willingness to pay in the two groups; thus, giving information about the harmful effects of chemical pesticides and visiting the control farms can be effective. The income factor is insignificant in owners' equation and has less effect in non-owners; it shows that the two groups are unaware of the benefits of such operation; hence, by raising awareness of the utility and demand of organic agricultural products, the policymakers can encourage the farmers to reduce the pesticides use. Owners have more motivation for this kind of operation, because they can utilize

S. Abedi (✉) · P. Bagheri · E. Pishbahar
Department of Agricultural Economics, University of Tabriz, Tabriz, Iran
e-mail: sahar_abedi91@yahoo.com

P. Bagheri
e-mail: pariya.ba93@gmail.com

E. Pishbahar
e-mail: pishbahar@yahoo.com

M. Rashidghalam (ed.), *The Economics of Agriculture and Natural Resources*,
Perspectives on Development in the Middle East and North Africa (MENA) Region,
https://doi.org/10.1007/978-981-15-5250-2_7

the long-run benefit. Hence, legislation about long-run rental contract and utilization of tax punishment for excessive use of pesticides can encourage non-owner to implement this operation.

Keywords Distribution of residuals · Endogenous switching copula · Khuzestan province · Ownership · Willingness to pay

1 Introduction

In agricultural sector, pests and insects are main sources of yield and quality reduction, and have *negative effects* on biodiversity and human health (Yazdi et al. 2010). Due to the specialization of agricultural activities and the increasing demand for food, pesticides are widely used. Excessive and unbalanced use of chemical fertilizers pollutes water and soil, endangers human and animal health, and increases production costs and dependency on external inputs. Therefore, it is necessary to consider the sustainable agriculture in response to the environmental and economic effects of conventional agriculture (Rasul and Thapa 2004). One of the most important operations of sustainable agriculture is Integrated Pest Management (IPM) which applies all compatible methods in a unit program to control pest populations and to prevent economic losses. According to the Food and Agriculture Organization (FAO) definition, the IPM is a precise investigation of all pest control techniques and their subsequent integration to prevent pest population growth so that the operations must be economical and have the least risks to human health and environment. IPM emphasizes on the growth of healthy crop production with the least disruption to ecosystems, and supports natural pest control techniques (FAO 2017). IPM includes three important methods: 1. crops improvement methods (plow, soil disinfection and replacement, use of resistant varieties, weed control, soil temperature–humidity–salinity control, and optimal plant nutrition); 2. Biological control (parasitoids, pathogens, and predators); 3. Chemical control (use of *least risk pesticide*s with short preharvest interval (PHI)[1], spray after harvesting, and do not use systemic pesticides). Optimal use of such methods controls pest population and reduces the environmental effects of fertilizers and pesticides (Farid et al. 2015); 4. Mechanical method (use sunlight, heat cold, and wind). Its components are shown in Fig. 1.

The IPM techniques first emphasize on monitoring pest population and use of pesticide only when economic threshold is exceeded, then focus on resistant varieties, biological control, and crops improvement methods application. Nowadays, IPM emphasizes on managing of host stress rather than perishing pests to increase plant tolerance. In fact, this method focuses on managing than controlling, and integrates multiple techniques (Alwang et al. 2019).

According to the Ministry of Agriculture Jihad (2018), about 12 and 2.5 million ha are under the chemical and non-chemical control, respectively, so that non-chemical

[1]The preharvest interval (PHI) is the wait time that must lapse after a pesticide application to a crop ready to be harvested (National Pesticide Information Center (NPIC) 2019).

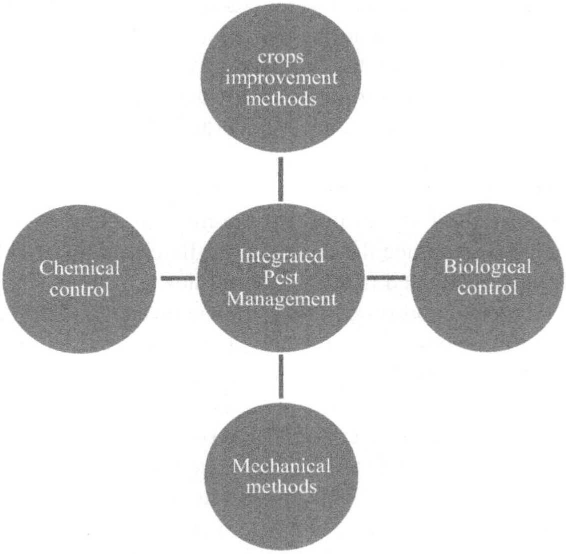

Fig. 1 Integrated pest management components. *Source* Pishbahar et al. (2019)

control is too low. In this regard, the amount of chemical pesticides use in Iran is 1,665,588 tons that 12.91% are used in Khuzestan[2] and this province is accounted for the greatest consumer of chemical pesticides (Ministry of Agriculture Jihad 2018). Pimentel (1991), Gajanana et al. (2006), Wandji et al. (2006), Leanne et al. (2008), Dhawan et al. (2009), Pretty and Bharucha (2015), Zahangeer Alam et al. (2016), Gautam et al. (2017), and Rahman et al. (2018) showed that the application of IPM operations can reduce costs and improve revenue and profit. Therefore, according to the definition of IPM and studies, the application of IPM operations in Khuzestan Province can be an effective step in reducing the amount of chemical pesticides, prevent their adverse effects, reduce costs, and improve revenue and profit.

Although, Khuzestan Province is ranked first in chemical pesticides use, according to the surveys conducted in different areas, farmers usually use integrated techniques, including, crop rotation, resistant varieties, useful insects, optimization of chemical fertilizer and pesticides consumption, improved seeds, and irrigation management. Various factors are effective on the application of these techniques; one of the important variables is ownership that has been used in different studies about the effective factors on willingness to pay and farmer's adoption of IPM. Neill and Lee (1999), Beckmann and Wesseler (2003), Surangsri et al. (2005), Anonymous (2007), Birungi (2007), Tatlidil et al. (2009), Hossein Zad et al. (2010), Carlberg et al. (2012), Gallardo and Wang (2013), Humayun Kabir and Rainis (2015), Adeli et al. (2017), and Babendreier et al. (2019) pointed that the ownership had a positive effect on the

[2]Khuzestan Province is one of the most important agricultural production areas in Iran. It is in the southwest of the country, bordering Iraq and the Persian Gulf and has the highest cereal production (13.12%) in the country (Ministry of Agriculture Jihad 2018).

farmer's adoption of IPM and increased their willingness to pay for IPM operations. They concluded that landowners accessed to more sources, had more motivation, and lots of crop cultivation methods choices, willingness to implement international agreements, and adoption of new marketing practices. In contrast, non-owners faced with financial constraints, concerned about less profit with IPM activities and also considered agriculture as a short-term activity. Hence, considering the differences between owner and non-owner farmers' conditions, it is expected that they have two different sets of priorities in implementing of IPM operation and pay different amount to reduce the chemical fertilizers and pesticides risks. However, this hasn't been investigated in any of the studies.

The main purpose of this study is investigation of factors affecting on the amount of willingness to pay in Khuzestan Province to implement IPM operations by farmland owners and non-owners. Regarding the importance of sustainable agriculture, studies in this field must be practical and accurate. Ownership factor and differences between the economic and social characteristics of the two groups make them respond differently to sustainable agricultural practices. Therefore, it is necessary to use more precise models to separate individuals. In this regard, to study the effective factors on willingness to pay to reduce environmental risks of pesticides for the two groups in Khuzestan Province the endogenous switching method with selection of best marginal distribution is applied to reach better results. It is hoped that the results of this study be an effective step in promoting sustainable agriculture planning.

2 Methodology

The copula and its theorem were first introduced by Sklar (1959). In recent years, this approach has been used in various economic fields and has improved the econometrics models. A copula is a function that connects the marginal distributions to obtain a joint distribution. In other words, a copula function $C : [0, 1]^2 \rightarrow [0, 1]$ has the following properties (Wichian and Sriboonchitta 2014):

1. For (u,v) in [0,1], $C(u, 0) = 0 = C(0, v)$, $C(u, 1) = u$, and $C(1, v) = v$
2. For every u_1, u_2, v_1, v_2 in [0,1] so that $u_1 \leq u_2$, and $v_1 \leq v_2$, $C(u_2, v_2) - C(u_2, v_1) - C(u_1, v_2) + C(u_1, v_1) \geq 0$

Sklar's theorem: Consider two random variables X and Y, with marginal distributions F and G, respectively, and a joint distribution H. Then, there is a copula function C for all x and y in $\overline{R}$ as follows:

$$H(x, y) = C(F(x), G(y)) \tag{1}$$

If marginal distributions F and G are continuous, C is unique, and otherwise, it is uniquely obtained on $\text{Ran}F \times \text{Ran}G$. The Eq. (1) is conversely available, i.e., if C is a copula function and F and G are marginal distributions, then H can be defined. Let $x = F^{-1}(u)$ and $y = G^{-1}(v)$ be the unique inverse transformations, then the copula can be defined as:

$$C(u, v) = H(F^{-1}(u), G^{-1}(v); \theta) \tag{2}$$

where u and v denote the standard uniform data. θ is the copula parameter, so-called dependency parameter. The copula parameter measures the dependency between two marginal distributions, according to the type of family, has different characteristics (Trivedi and Zimmer 2005).

Although θ indicates the degree of dependency, it is not directly possible to compare the parameter of a copula with other copulas parameters, because its nature varies in different copulas. Therefore, we can use the concordance criteria, such as Kendal' τ which is applied in many statistical fields. The Kendal' τ can be defined in terms of copula as follows (Nelson 1998):

$$\tau = 4\int\int_{[0,1]^2} C_\theta(u, v)dC_\theta(u, v) - 1 = 4E(C_\theta(U, V)) - 1 \tag{3}$$

where E is the expected value of $C_\theta(U, V)$. The bound of Kendal' τ is between −1 and 1, and if the variables are independent, then $\tau = 0$ and the product copula is suitable (Wichian and Sriboonchitta 2014). If the value of Kendal' τ is closer to $|1|$, a stronger dependency exists. The interesting point is that Kendal' τ can be expressed in term of copula parameter. In sample selection models, the independent test of the error components is important. In the case of independence, it is better to use OLS to estimate the parameters. Therefore, besides Kendal' τ, we can use likelihood-ratio test, which is asymptotically distributed as χ^2, to verify the null hypothesis of independence (Hasebe 2013).

There are different families of copulas, which differ in the form of function, properties, and shapes of distribution, like skewness to left or right, symmetric and asymmetric, thin or fat tails. Gaussian or normal copula, introduced by Lee (1983) for the selectivity models, is a comprehensive function, includes the product one and both of the Frechet–Hoeffding bounds. It also has a symmetrical and a strong central dependency structure, and its dependency parameter is between −1 and 1, and $-1 \leq \tau \leq 1$. The other family is Farlie–Gumbel–Morgenstern (FGM). Although this copula is symmetric and similar to Gaussian copula, its dependency structure is weaker. In addition, this copula is not comprehensive and its Kendal' τ domain is limited to $-2/9 \leq \tau \leq 2/9$. Besides these copulas, the *Plackett* copula is also widely used in literature, which is symmetric and has the same dependency in upper and lower tails (Trivedi and Zimmer 2005).

There is an important class known as Archimedean copulas that are widely used in empirical studies. These families can show a wide range of dependencies using different generator functions. Smith (2003) also showed that these functions simplify the estimation of maximum likelihood and the score function. The bivariate Archimedean copula is defined as:

$$C_\theta(u, v) = \phi^{-1}(\phi(u) + \phi(v)) \tag{4}$$

where $\phi : [0, 1] \rightarrow [0, \alpha]$ denotes a generator function with following properties: $\phi(1) = 0$, $\phi'(t) < 0$ and $\phi''(t) > 0$ for all $0 < t < 1$. Moreover, if $\phi(0) = \alpha$, then this function can be inverted and ϕ^{-1} is its inverse function.

The Ali–Mikhail–Hagh (AMH), Clayton, Frank, Gumbel, and Joe are Archimedean copulas that are widely applied in literature. These copulas have different generator functions which lead to different functional forms and dependency structures. The Frank copula is comprehensive and symmetric, has a central dependency, and $-1 \leq \tau \leq 1$. In contrast, the AMH, Clayton, Gumbel, and Joe are not symmetric and comprehensive. The Clayton, Gumbel, and Joe copulas only allow for positive dependency due to $0 \leq \tau < 1$. It seems to be a limitation, while we can easily solve it using the negative form of the corresponding variable. This modification does not change the model structural information and allows for negative dependency between the variables with copulas in $-1 < \tau \leq 0$ (Hasebe and Vijverberg 2012).

In general, an endogenous switching regression can separate a model into two regimes: the employment in private or public sector, union or non-union status, ownership and non-ownership, etc. In this study, according to two regimes, i.e., ownership and non-ownership, the effective factors on willingness to pay are investigated. This issue can be explained by the system of equations as:

$$\text{WTP}_{1i} = X_{1i}\beta_1 + \varepsilon_{1i} \quad \text{if} \quad S_i = 1 \quad \text{or} \quad Z_i\gamma + \varepsilon_{si} > 0 \tag{5}$$

$$\text{WTP}_{0i} = X_{0i}\beta_0 + \varepsilon_{0i} \quad \text{if} \quad S_i = 0 \quad \text{or} \quad Z_i\gamma + \varepsilon_{si} \leq 0 \tag{6}$$

$$S_i^* = Z_i\gamma + \varepsilon_{\text{si}}u \tag{7}$$

where WTP_{1i} and WTP_{0i} are the amount of willingness to pay of individual i be owner and non-owner, respectively, X_{1i} and X_{0i} are the vectors of independent variables that affect the amount of willingness to pay. Equation (7) is switching equation. S_i^* represents the latent or decision variable, as to whether the individual i is owner ($S_i = 0$) or non-owner ($S_i = 1$), and is a function of the vector of individual characteristics, Z_i that influence the individual's decision on ownership. If $S_i^* = 1$, then WTP_{1i} is available, and WTP_{0i} is available if $S_i^* = 1$. β_1, β_0, and γ represent the vectors of parameters, ε_1, ε_0, and ε_s are error terms which have a normal distribution with mean vector of 0 and covariance matrix as follows (Wichian and Sriboonchitta 2014):

$$\Omega = \begin{bmatrix} \sigma_{11} & \sigma_{10} & \sigma_{1s} \\ \sigma_{01} & \sigma_{00} & \sigma_{0s} \\ \sigma_{s1} & \sigma_{s0} & 1 \end{bmatrix} \tag{8}$$

In endogenous switching regression models, it is important to note if there is a significant difference between the amounts of willingness to pay by two regimes.

The answer determines the correction of model separation. In this case, the average treatment effect (ATE) is calculated and, if this effect is zero, then there is no difference between the amounts of willingness to pay by two groups and the endogenous switching is not a suitable model. Thus, the ATE is calculated as:

$$E(W\hat{T}P_1 - W\hat{T}P_0|X_1, X_0) = X_1\hat{\beta}_1 - X_0\hat{\beta}_0 \tag{9}$$

where $W\hat{T}P_i$ is the estimated values of willingness to pay by two groups according to the estimated coefficients, $\hat{\beta}_i$.

If ε_1, ε_0, and ε_s are not independent, the OLS regression leads to inconsistent estimation of parameters. Therefore, the full information maximum likelihood (FIML) method is used to estimate these equations. The general form of the function of likelihood in the endogenous switching regression model is presented as (Hasebe 2013):

$$L = \prod_0 \int_{-\infty}^{-z'\gamma} f_{s0}(\varepsilon_s, \varepsilon_0)d\varepsilon_s \prod_1 \int_{-z'\gamma}^{\infty} f_{s1}(\varepsilon_s, \varepsilon_1)d\varepsilon_s \tag{10}$$

where f_{s0} is the joint density function of ε_s and ε_0, and f_{s1} is the joint density function of ε_s and ε_1. To estimate Eq. (10), the distribution of density functions have to be determined. For this, we can use normal distribution, while this wrong assumption leads to inconsistency in estimation (Hasebe and Vijverberg 2012).

In order to solve this issue in sample selection models, Lee (1983) proposed copula approach that expanded by Smith (2003). Hence, Eq. (10) can be rewritten as:

$$L = \prod_0 \left\{ \frac{\eta'(F_0)}{\eta'(C_\lambda(F_s, F_0))} \right\} f_0 \prod_1 \left\{ 1 - \frac{\phi'(F_1)}{\phi'(C_\theta(F_s, F_1))} \right\} f_1 \tag{11}$$

where $\eta' = \partial\eta(t)/\partial(t)$ and $\phi' = \partial\phi(t)/\partial(t)$. F is the cumulative distribution function, η and ϕ are copula functions, λ and θ are the dependency parameters of ε_1, ε_0, and ε_s, and $F_s = F_s(0)$, $F_0 = F_0(\varepsilon_0)$, $F_1 = F_1(\varepsilon_1)$, $f_s = f_s(0)$, $f_0 = f_0(\varepsilon_0)$, and $f_1 = f_1(\varepsilon_1)$ (Wichian and Sriboonchitta 2014).

To perform the maximum likelihood estimation, the functional form of the marginal distributions of error terms and the dependency structure or suitable copula, that connects them, needs to be specified. If the dependency pattern is known, copula selection will be simple. However, such information is rarely available. The copula functions mentioned here are not nested so that AIC and BIC criteria can be used. Similarly, the marginal distributions have to be selected. In fact, the marginal distribution can be selected from various number of univariate distributions which is the advantage of copula method. We can select normal and logistic distributions that are corresponding to the probit and logit models to explain the switching equation. Normal, logistic, and Student's *t*-distributions can also be considered for willingness to pay equations. Among these marginal distributions, Student's *t*-distribution is more flexible. When the degrees of freedom of Student's t go to infinity, its shape approximates the normal distribution. In addition, when the degree of freedom is about 8, Student's t approximates the logistic distribution. The smaller degrees of freedom of Student's t can show thicker tails than others. For this, Hasebe (2013) recommended the Student's t-distribution. For the switching equation, the type of distribution has no significant effect on results since probit and logit models do not have a significant different in a binary choice model; however, the Vuong test can be useful in choosing between normal and logistic distributions. This test compares two nested statistical models, and is LR-based that measures the distance between two models. Following the literature, the empirical endogenous switching copula models are presented as follows:

$$\mathrm{OWN} = \gamma_0 + \gamma_1\mathrm{FL} + \gamma_2\mathrm{COST} + \gamma_3\mathrm{DIST} + \gamma_4\mathrm{AGE} + \gamma_5\mathrm{WAGE} + \gamma_6\mathrm{OJOB} \tag{12}$$

$$\mathrm{WTP}_1 = \beta_{01} + \beta_{11}\mathrm{COST}_1 + \beta_{12}\mathrm{KNOW}_1 + \beta_{13}\mathrm{REV}_1 + \beta_{14}\mathrm{FSZ}_1 + \beta_{15}\mathrm{SEV}_1 \quad \text{if} \quad \mathrm{OWN} = 1 \tag{13}$$

$$\mathrm{WTP}_0 = \beta_{00} + \beta_{01}\mathrm{COST}_0 + \beta_{02}\mathrm{KNOW}_0 + \beta_{03}\mathrm{REV}_0 + \beta_{04}\mathrm{FSZ}_0 + \beta_{05}\mathrm{SEV}_0 \quad \text{if} \quad \mathrm{OWN} = 0 \tag{14}$$

where OWN: a dummy variable of landownership (private = 1, otherwise = 0) as a dependent variable, and FL: the family member labor (people), COST: the total variable cost (million Rial), DIST: the distance from the road (km), AGE: the age of farmer (year), WAGE: the daily labor wage (10,000 rial) and OJOB: the income from non-agricultural activities (have another income = 1 and have no income from another job = 0) as independent variables in the switching equation.

WTP is the weighted average of the willingness to pay in two groups (owner = 1 and non-owner = 0) and three levels of risk (10,000 rial). To calculate WTP, the environment is divided into five layers (humans, mammals, birds, aquatic species, and beneficial insects). In addition, the pesticides' risk is classified into three risk levels ($j = 1, 2, 3$) for each environmental layer ($i = 1,\ldots, 5$), i.e., there are $5 \times 3 = 15$ classes of environmental layers/risk levels for each pesticide. Then, the willingness to pay to avoid negative consequences of pesticide on the environment and human health is measured. For this, the farmers are asked how much they would pay to avoid such risks; hence, the willingness to pay for each class is measured as:

$$\mathrm{WTP}_{\mathrm{ih}} = \frac{\mathrm{Importance}_i}{\sum_{i=1}^{5} \mathrm{Importance}_i} \times \mathrm{WTP}_j \tag{15}$$

where WTP_j is the individual willingness to pay to reduce three levels of risk (j) in five environmental layers, and $\mathrm{Importance}_i$ is the level of importance to prevent a given risk levels associated with each environmental layer (i). The value of three WTP_j in five layers is not equal, because it is expected that the reduction in consumption of pesticide which has the highest level of risk for humans will be more valued than the same reduction in the pesticide that has the highest risk for other layers. In order to properly allocate the willingness to pay to each class, an appropriate indicator should be defined to distribute its values based on the level of importance. For this purpose, the average of willingness to pay is calculated for all three levels. The weighted difference between the distance of low and moderate risk averages with the distance of high and moderate risk average forms the indicator range based on the level of importance. The independent variables include COST: the total variable cost (million Rial), KNOW: the sum of farmer's knowledge about the pesticides' environmental risks to each layer (0 = no knowledge, 1 = knowledge), REV: the total income (million rial), FSZ: the cultivated area (ha) and SEV: pest density (low density = 1, moderate density = 2, and high density = 3). Using the two-stage sampling method, the required data were collected from 164 farmers (96 farmers were owner, and 68 farmers were non-owners) of Khuzestan Province in 2017. To estimate the model, we used the Stata software.

3 Results and Discussion

The *socioeconomic* characteristics of owners and non-owners are presented in Table 1. This table shows that 58.54% of farmers are owners. The averages of landowners and non-owners age are 50.56 and 37.45 years, respectively; it indicates that owners are older than non-owners. The variable of production costs for non-owner and owner farmers are 1530 and 1150 million rial, respectively. Because of land rent, such costs are higher for non-owner. The averages of non-owners and owners' income are 3830 and 5140 million rial. About 25% of owners and 14.7% of non-owners do other businesses besides agriculture. 66.7% of farmers are aware of IPM programs so that 83.3% of them consider these programs importance. 23.89% of farmers participate in the classes of school in field, and consequently, by improving pest and disease management, they reduce the consumption of pesticides, on average, 52.27%. Although, owners and non-owners awareness of chemical pesticides harmful effects are almost the same, owners pay more attention to such effects so that the average cost of pesticide per hectare for owner is 140 million rial and for non-owners is 160 million rial.

Table 1 Characteristics of farmers in Khuzestan province

Variables	Non-owners (41.46)	Owners (58.54)
Average age (year)	37.45	50.56
Average cost (million rial)	1530	1150
Average income (million rial)	3830	5140
Other business (%)	14.7	25
Average cost of pesticide per hectare (million rial)	160	140
The family member labor (people)	1.56	1.68

The average of willingness to pay for owners and non-owners in each environmental class (environmental layers/risk levels) are reported in Tables 2 and 3, respectively. For owners, the highest average of willingness to pay to reduce the harmful effects of pesticides on the environmental layers is evident for humans, mammals, beneficial insects, birds, and aquatic species, respectively. Similarly, for non-owners, the highest average is corresponding to humans, and then with mammals, birds, beneficial insects, and aquatic species. The reduction in consumption of pesticide, which has the highest level of risk for humans, is more valuable than the same decline for others layers; hence, the highest amount of willing to pay in the two groups is corresponding to human layer. According to Tables 2 and 3, it is also clear that owners are willing

Table 2 Average willingness to pay of owners in each environmental class

Risk level				
Environmental layers	High risk = 3	Moderate risk = 2	Low risk = 1	WTP average of risk levels for each layer
Humans = 1	357,929.66	234,935.26	122,311.29	23,8392.07
Mammals = 2	125,235.74	83,125.67	44,078.199	84,146.536
Birds = 3	98,885.148	64,524.34	35,217.674	66,209.054
Aquatic species = 4	117,061.32	76,608.718	39,210.919	77,626.98
Beneficial insects = 5	148,856.89	97,681.016	53,296.502	99,944.801
WTP average of layers for each risk level	169,593.75	111,375	58,822.917	113,263.89

Table 3 Average willingness to pay of non-owners in each environmental class

Risk level				
Environmental layers	High risk = 3	Moderate risk = 2	Low risk = 1	WTP average of risk levels for each layer
Humans = 1	251,247.690	151,131.040	81,349.597	161,242.77
Mammals = 2	78,108.079	49,563.417	29,380.187	52,350.561
Birds = 3	63,378.133	41,527.419	25,623.711	43,509.754
Aquatic species = 4	83,657.354	51,756.226	30,451.664	55,288.41
Beneficial insects = 5	99,711.69	59,477.782	32,900.723	64,030.06
WTP average of layers for each risk level	115,220.59	70,691.176	39,941.176	75,284.314

to pay more to reduce the harmful effects of pesticides; therefore, owners value more for such programs.

In order to investigate the effect of variables on decision or switching variable and the amount of willingness to pay to reduce the harmful effects of pesticides on the environmental layers in the two groups, the endogenous switching copula method is used to explain the system of equations. The Student's t-distribution is selected for error terms in willingness to pay equations as mentioned above, it includes both distributions. However, the proper distribution has to be selected for error term in switching equation. Furthermore, the appropriate copula must be selected to determine the joint distribution of error terms in the system of equations. In this study, we investigate nine copula functions, i.e., product, Gaussian, FGM, *Plackett*, AMH, Clayton, Gumbel, Frank, and Joe. For this purpose, considering the normal (probit) and the logistic (logit) distribution for error term in the decision equation, 81 different models for each pattern (162 models) are investigated, and the appropriate copula is selected based on the smallest amount of AIC and BIC. Thus, the combination of product–Gaussian copula with AIC = 1180 and BIC = 1233.27 for probit-t-t pattern and AMH–Gaussian copula with AIC = 1179.44 and BIC = 1234.93 for logit-t-t pattern are selected. After selecting appropriate copulas, the probit-t-t and logit-t-t patterns are estimated using the endogenous switching copula method. The Vuong test is used to select the appropriate pattern, and to compare the log-likelihood in the two patterns. The Vuong statistic is equal to −4.65 and the p-value is 0; therefore, the null hypothesis or the absence of a significant difference between the log-likelihood of the two patterns is rejected at all significant levels. Consequently, the logit-t-t pattern is statistically more appropriate than the probit-t-t to examine the affecting variables on the system of equations. The result of the logit-t-t pattern is reported in Table 4.

In Table 4, the atheta0 and atheta1 represent the ancillary dependency parameters that can be transformed to the AMH and Gaussian copulas parameters. The AMH parameter is 0.83 and the Gaussian parameter, which is the linear correlation coefficient, is 0.93; since these parameters cannot be interpreted, the Kendal' τ is calculated. This statistic for error terms correlation in the decision and the owner equations is −0.25 and in the switching and the non-owner equations is −0.76, which rejects the assumption of independence of errors in the switching and the willingness to pay equations. The negative value of this correlation coefficient indicates a negative relationship between the equations' error terms. The lndf0 and lndf1 parameters are also the ancillary parameter of degrees of freedom of Student's t. The degree of freedom of Student's t (df) for owners equation is equal to 1.62 and for non-owners is 5.21. These estimated amounts show that the distributions of error terms in willingness to pay equations have much thicker tails than normal; therefore, the Student's t-distribution for the error terms is suitable. Furthermore, since the degree of freedom of the non-owners equation is greater than the owners, the error terms distribution in the owners' equation is thicker. The lnsigma0 and lnsigma1 statistics are also the ancillary parameter to calculate the scale parameter and standard deviation of error terms. Besides the Kendal' τ, the LR statistic is also used to examine the independence of the error terms. The LR statistic is 8.749 that rejects the independence of

Table 4 Results of endogenous switching copula model based on the logit-t-t model

Variables	Switching equation				Willingness to pay equations			
	Logit				Owner		Non-owner	
	Coeff	Z	Marginal effect	Elasticities	Coeff	Z	Coeff	Z
Cons	−5.51	−2.51*			−2319.65	−2.21**	−1662.13	−2.62*
FL	−0.39	−1.29***	−0.08	−0.208				
COST	0.01	1.94**	0.003	0.830	0.00008	2.22**	0.00002	0.99
DIST	−0.15	−1.80***	−0.034	−0.114				
AGE	0.08	2.48*	0.018	1.069				
WAGE	0.04	0.54	0.009	0.093				
OJOB	3.84	2.95*	0.472	0.213				
FSZ					72.87	1.74***	−0.001	−3.96*
KNOW					321.40	3.96*	215.65	4.50*
REV					−2/07	−0.14	0.00001	1.72***
SEV					778.14	3.97*	159.32	1.31
Lnsigma0	6.59	51.41*						
Lnsigma1	6.28	26.64*						
Lndf0	1.482	1.04*						
Lndf1	1.64	2.07**						
Atheta0	1.18	1.43***						
Atheta1	1.66	1.50***						
sigma0	729.58							
sigma1	534.88							

(continued)

Table 4 (continued)

Variables	Switching equation				Willingness to pay equations			
	Logit				Owner		Non-owner	
	Coeff	*Z*	Marginal effect	Elasticities	Coeff	*Z*	Coeff	*Z*
theta0	0.82							
theta1	0.93							
df0	1.62							
df1	5.20							
Tau0	−0.246							
Tau1	−0.762							
LR			8.749		Log-likelihood		−564.71	
P-value			0.0126		ATE		12.52 million rial	

*, **, and *** are significant level at 1, 5, and 10%, respectively

error terms, and shows that the OLS estimation leads to inconsistent results. The ATE is equal to 12.52 million rial; this value indicates that there is a significant difference between the willingness to pays of the two groups, so that the owners tend to pay more to reduce the harmful effects of pesticides on the environmental layers. These results are in line with the studies by Neill and Lee (1999), Beckmann and Wesseler (2003), Surangsri et al. (2005), Anonymous (2007), Birungi (2007), Tatlidil et al. (2009), Hossein Zad et al. (2010), Carlberg et al. (2012), Gallardo and Wang (2013), Humayun Kabir and Rainis (2015), Adeli et al. (2017), and Babendreier et al. (2019). They have demonstrated that the ownership has a positive effect on the farmer's adoption of IPM, we can conclude that the separation of the willingness-to-pay equation into two categories and the use of endogenous switching model is correct.

In the switching equation, the number of family labor and the daily labor wage are not significant. The distance from the road has a significant negative effect, while the total variable cost, the farmer age, and the income from non-agricultural activities have a positive and significant effect on the probability of ownership decision. Since the estimated coefficients of the logit-t-t model are not interpretable, the marginal effects and their elasticity are also calculated. The results of the marginal effects indicate that the income from non-agricultural activities with 0.47 has the greatest effect on the ownership decision probability. The marginal effect of this variable shows that given the other factors, if a person has income from non-agricultural activities, the probability of ownership decision increases by 47%.

According to the elasticity of variables, the rank of effecting percentages on the probability of ownership decision are corresponding to farmer age, total variable cost, incomes from non-agricultural activities, and distance from the road. As mentioned before, the distance from the road has a negative effect on the probability of decision-making. This variable elasticity is −0.114 and shows that given the other factors, for each added percentage in the distance between the agricultural land and the road, the ownership decision probability reduces by 11.4 percent. The age variable shows that older farmers with more affordability are more likely to own than young farmers. Increasing variable costs reduces the tendency to produce so that the farmer prefers to invest in fixed asset like land. Having income from non-agricultural activities also increases the individual's financial ability and the probability of ownership by 21.3%. The results of the willingness to pay equations for the two groups are also reported in Table 4. The estimated coefficient of income for owners and the total variable cost and the pest density for non-owners are not significant. All significant variables have positive effects in owners' equation, while in non-owners equation the cultivated area has a negative effect on the amount of willingness to pay. Since one of the most important factors in the variable cost is the pesticides cost, increase in such costs makes the owners interested in IPM and tend to pay more. Increase in owners' cultivated area increases the amount of willingness to pay to 728.7 thousands rial, while for non-owners reduces the willingness to pay. The reason is that by increasing the cultivation area, the cost of land increases as well, and it is not economical to use methods with long-run impact on pest control so that non-owners prefer to use short-run methods such as chemical pesticides. One percent increase in the knowledge level increases the owners and non-owners willingness to pay to 3214 and 2156.5 thousands rial, respectively. The positive and significant impact of this variable indicates that farmers who are more familiar with the hazards of chemical pesticides, and the effectiveness of the IPM in reducing environmental risks and production costs; hence, they are more likely to accept these operations. In addition, the averages of owners and non-owners level of knowledge about such issue and their level of education are almost the same; however, the effect of knowledge in owners' equation is more than non-owners, that indicates the effect of ownership. Although, the total income is significant in non-owners equation, its effect on the amount of willingness to pay is too small. The insignificance of this variable in owners' equation can shows that the owners, regardless of their income from agricultural activities, have more willingness to pay to reduce the negative environmental effect of chemical pesticides. In fact, the ownership variable has a tremendous effect on the prospect of future profits of the integrated pest management operations. One percent increase in the pest density increases the owners' willingness to pay to 7781.4 thousands rial, but as mentioned, this variable is not significant in non-owners equation. Landowners can benefit from the long-run effects of the IPM, while this is not true about the non-owners, therefore they do not pay attention to the severity of pest density in the amount of willingness to pay. It should be noted that the owners and non-owners reside in the studied region.

4 Conclusion

The difference between the owner and non-owner farmers' characteristics makes them have different priorities in the use of IPM; therefore, it is expected that the amount of willingness to pay and its affecting factors are different in the two groups. In such situation, the selection is done endogenously; hence, the endogenous switching models results better than the other sample selection models. The main problem that is presented in these models is the system of equations estimation and the correlation between the error terms in the switching and the willingness to pay equations. The common solution is the multivariate normal distribution assumption in the full information maximum likelihood estimation method. However, in the real world, this may lead to incorrect results due to the selectivity bias. Therefore, in this study, we used copula approach to cope with such wrong assumption. The copula approach allows the marginal distribution to be selected from various numbers of flexible distributions, thus leads to efficient and the accurate estimation. Therefore, according to the Khuzestan Province rank in terms of chemical pesticides, the affecting factors on the farmers' ownership and the amount of willingness to pay to reduce the environmental hazards of chemical pesticides in the two groups were investigated using the endogenous switching copula method.

The results showed that due to the small value of degrees of freedom, the Student's t-distribution is desirable for the error terms, and their distributions in the willingness to pay equations have thicker tails than the normal distribution. In addition, the Vuong test showed that the logistic distribution was better than the normal distribution for the decision equation. Both of these results provide a good proof about the ineffectiveness of the normal assumption in the maximum likelihood estimation of the endogenous switching method; therefore, the copula method leads to more flexibility in the dependency structure investigation. Furthermore, it is recommended using more flexible distributions that are more consistent with the error terms marginal distributions to increase the result accuracy.

The results of the ATE showed that there was a significant difference between the amounts of willingness to pay in the two groups. Consequently, as the willingness to pay equations results showed, the effective factors were different in the two groups. Landowners were more concerned about the soil pollution and the environmental hazards because the land was in their own possession, as it was evident in more willingness to pay and lower costs associated with pesticides. The knowledge level was one of the most important factors that had a positive effect on the willingness to pay to reduce the risks of pesticides in the two groups; thus, providing posters and programs that depict the harmful effects of chemical pesticides, such as vegetation damages, soil degradation, and water pollution, as well as visiting the control farms (that use chemical pesticides and produce organic crops) can be effective tools to reduce the consumption of these materials. The insignificant effect of income in owners' equation and the less significant effect of it in non-owners indicate that the two groups are unaware of the benefit of IPM, hence by raising awareness of the utility and demand of organic agricultural products, and incentives tools like the higher price determination for such products, the policymakers can encourage the farmers to reduce the consumption of chemical pesticides. Due to the fact that the owners can use the long-run benefits of the IPM operations, they have more willingness to pay than non-owners; therefore, the rules for long-run contracts may solve this problem. In addition, using the penalties for more usages can reduce the consumption of pesticides and increase the use of IPM.

References

Adeli M, Khodaverdizadeh M, Hayati B (2017) Application of ordered logit model to determine factors affecting adoption of integrated pest management practices among greenhouse owners in Jiroft county. J Sci Technol Greenhouse Cult 8(3):107–119

Alwang J, Norton G, Larochelle C (2019) Obstacles to widespread diffusion of IPM in developing countries: lessons from the field. J Integr Pest Manage 10(1):1–8

Anonymous (2007) Point of views agricultural organizations in reducing consume of pesticide and implementation biological control project. Reported the Second Gathering of Head of Agricultural Organizations Provinces of Iran

Babendreier D, Wan M, Tang R, Gu R, Tambo J, Liu Z, Grossrieder M, Kansiime M, Wood A, Zhang F, Romney D (2019) Impact assessment of biological control-based integrated pest management in rice and maize in the greater Mekong subregion. Insects 10(226):2–16

Beckmann V, Wesseler J (2003) How labor organization may affect technology adoption: an analytical framework analyzing the case of integrated pest management. Environ Dev Econ 8(3):437–450

Birungi PB (2007) The linkage between land degradation, poverty and social capital in Uganda. Ph.D. Thesis, Faculty of Agricultural Economics, Extension and Rural Development, University of Pretoria, South Africa

Carlberg E, Kostandini G, Dankyi A (2012) The effects of integrated pest management techniques (IPM) farmer field schools on groundnut productivity: evidence from Ghana. Selected Paper Prepared for Presentation at the Agricultural and Applied Economics Association

Dhawan AK, Singh S, Kumar S (2009) Integrated pest management (IPM) helps reduce pesticide load in cotton. J Agric Sci Technol 11(5):599–611

Farid S, Pourkhaton M, Lori Z (2015) Integrated pest management in greenhouses. Ministry of Agriculture-Jahad Agriculture Organization of Kerman Province, Kerman Agricultural Promotion Coordination Management, pp 1–16

Food and Agriculture Organization of the United Nations (2017) Integrated pest management of major pests and diseases in Eastern Europe and the Caucasus

Gajanana TM, Krishna Moorthy PN, Anupama HL, Raghunatha R, Prasanna Kumar GT (2006) Integrated pest and disease management in tomato: an economic analysis. Agric Econ Res Rev 19:269–280

Gallardo R, Wang Q (2013) Willingness to pay for pesticides' environmental features and social desirability bias: the case of apple and pear growers. J Agric Resource Econ 38(1):124–139

Gautam S, Schreinemachers P, Uddin MN, Srinivasan R (2017) Impact of training vegetable farmers in Bangladesh in integrated pest management (IPM). Crop Protect 102:161–169

Hasebe T (2013) Copula-based maximum-likelihood estimation of sample-selection models. Stata J 13(3):547–573

Hasebe T, Vijverberg W (2012) A Flexible sample selection model: a GTL-copula approach. IZA Discussion Paper No. 7003

Hossein Zad J, Shorafa S, Dashti GH, Hayati B, Kazemiyeh F (2010) An economic evaluation of the environmental benefits from pesticides reduction program in Khuzestan province. J Agric Sci Sustain 20(4):101–112

Humayun Kabir M, Rainis R (2015) Adoption and intensity of integrated pest management (IPM) vegetable farming in Bangladesh: an approach to sustainable agricultural development. Environ Dev Sustain 17(6):1413–1429

Leanne M, Mullen J, Stevens M (2008) An evaluation of the economic, environmental and social impact of investments in IPM research in invertebrate rice pests. Economic Research Report No. 41. https://www.ageconsearch.umn.edu

Lee LF (1983) Generalized econometric models with selectivity. Econometrical 51:507–512

Ministry of Agriculture-Jihad Statistic Information, Agricultural Statistical Yearbook (2018) Available at https://www.maj.ir

National pesticide information center (NPIC) (2019) Preharvest interval (PHI). Available at https://www.npic.orst.edu

Neill SP, Lee DR (1999) Explaining the adoption and disadoption of sustainable agriculture: the case of cover crops in northern Honduras. Department of agriculture resource and managerial economics, Cornell University working paper 31

Nelson RB (1998) An introduction to copulas. Lecture Notes in Statistics. Springer, New York

Pimentel D (1991) CRC handbook of pest management in agriculture. CRC Press, Boca Raton, p 784

Pishbahar E, Hosseinzad J, Abedi S, Bageri P (2019) Willingness to pay for IPM: an application of the Heckman-Copula approach. In: Rashidghalam M (eds) Sustainable agriculture and agribusiness in Iran. Perspectives on development in the Middle East and North Africa (MENA) Region. Springer, Singapore

Pretty J, Bharucha Z (2015) Integrated pest management for sustainable intensification of agriculture in Asia and Africa. Insects 6(1):152–182

Rahman M, Norton G, Rashid M (2018) Economic impacts of integrated pest management on vegetables production in Bangladesh. J Crop Protect 113:6–14

Rasul G, Thapa G (2004) Sustainability of ecological and conventional agricultural systems in Bangladesh: an assessment based on environmental, economic and social perspectives. Agric Syst 79:327–351

Smith MD (2003) Modelling sample selection using archimedean copulas. Econometrics J 6:99–123

Surangsri W, Suraphol C, Nunta B (2005) Factors affecting the adoption and non-adoption of sloping land conservation farming practices by small-scale farmers in Thailand. Report the Swiss Agency for Development and Cooperation (SDC) Bankok, Tailand

Tatlidil FF, Boz I, Tatlidil H (2009) Farmer's perception of sustainable agriculture and its determinants: a case study in Kahramanmaras province of Turkey. Environ Dev Sustain 11:1091–1106

Trivedi PK, Zimmer DM (2005) Copula modeling: an introduction for practitioners. Found Trends Econometrics 1(1): Now Publishers

Wandji N, Juliusa L, Gockowskia J, Isaacb T (2006) Socio-economic impact of a cocoa integrated crop and pest management diffusion knowledge through a farmer field school approach in southern Cameroon. In: Presentation at the international association of agricultural economists conference, Gold Coast, Australia

Wichian A, Sriboonchitta S (2014) Econometric analysis of private and public wage determination for older workers using a copula and switching regression. Thai J Math 111–128

Yazdi Z, Sarreshtedari M, Zohal MA (2010) Respiratory disease in workers exposed to organophosphate materials. J School Med 53(4):206–213

Zahangeer Alam M, Crump AR, Haque M, Islam S, Hossain E, Hasan S, Hasan SH, Hossain S (2016) Components in a rice agro-ecosystem in the Barisal Region of Bangladesh. Front Environ Sci 4:1–10

Sahar Abedi holds B.Sc. and M.Sc. in Agricultural Economics from Department of Agricultural Economics, University of Tabriz. For her M.A. Thesis, she aimed to have a better understanding of weather-based crop insurance premium for wheat crop. Her research interest lies at Risk Management and Crop insurance.

Pariya Bagheri holds B.Sc. in Agricultural Economics and M.Sc. in Production Economics in Department of Agricultural Economics, University of Tabriz. Her main research interest centers on Production Economics, Risk Management and Climate Change.

Esmaeil Pishbahar is an Associate Professor of Agricultural Economics at University of Tabriz, Iran. He holds a B.Sc. in Agricultural Economics from University of Tabriz and a M.Sc. in Agricultural Economics from University of Tehran. He did his Ph.D. in Science Economics at departments of Economics and Management, University of Rennes1, France. His areas of interest and research are Applied Econometrics, Agricultural Risk Management and Insurance, and International Trade. His teaching area are Advanced Econometrics, Mathematical Economics, and Macroeconomics at under- and postgraduate levels. He has over 100 publications in journals and chapters in books.

Estimating Economic Value and Compensation Surplus of Animal Species in Arasbaran Forests Using Choice Experiment Method

Maryam Haghjou, Babollah Hayati, Esmaeil Pishbahar, and Morteza Molaei

Abstract Sustainable development has no meaning without preserving forests, rangelands and natural resources. Regarding the necessity of protecting forests, improving forest management is one of the priorities of sustainable management. Applying of appropriate analytical tools and proper techniques of environmental valuation to concentrate in the issues of forests' functions from the economic point seems necessary to provide a context of sustainable development. Forests are great examples of environmental treasures to be evaluated in this route. Arasbaran is one of the conservational and protected forests of Iran, which due to its various biodiversity and the existence of more than 1072 species of plants and numerous animal species, is one of the nine Biosphere reserves in Iran. Therefore, the study aimed to estimate the economic value and compensation surplus (CS) of three valuable animal species of Arasbaran Forests: the bear, the tiger and the black rooster, using choice experiment method and application of conditional Logit regression model. Required data were acquired through field studies and questionnaires filled by 334 visitors and citizens from ten cities in three provinces: East Azerbaijan, West Azerbaijan and Ardabil. Results showed that total economic value of these three species is about 692.695×10^9 Rial (About 22×10^6 USD) and 50% of this value is allocated to the tiger. Also the results revealed that compensating surplus of relative and optimum improvement in animal species' condition would monthly worth about 3,913,333 and 50,733,333 Rial, respectively. Furthermore, results reveal that respondents' level of education, income, number of annual visits to the forests and their friendly attitudes towards Arasbaran Forests had significant positive impacts on Willingness to Pay of respondents for the animal species.

M. Haghjou (✉) · B. Hayati · E. Pishbahar · M. Molaei
Department of Agricultural Economics, University of Tabriz, Tabriz, Iran
e-mail: maryam.haghjou@yahoo.com

B. Hayati
e-mail: b-hayati@tabrizu.ac.ir

E. Pishbahar
e-mail: pishbahar@yahoo.com

M. Molaei
e-mail: morteza.molaei@gmail.com

M. Rashidghalam (ed.), *The Economics of Agriculture and Natural Resources*,
Perspectives on Development in the Middle East and North Africa (MENA) Region,
https://doi.org/10.1007/978-981-15-5250-2_8

Keywords Arasbaran forest · Choice experiment · Compensation surplus · Conditional logit model · Economic valuation of animal species

1 Introduction

The economic developments are globally speeding up in the recent decades and there are immoderate pressures on environment all around the world. Therefore, human being is witness of irreversible damage to natural resources in almost all countries specially the underdeveloped ones. Meanwhile, the principles of sustainable development encounter with destruction of natural resources and emphasis on keeping them for the future and respecting their right of being. Due to this fact, focusing on the negative impacts of development process is becoming an important issue; which leads the governments into a economic transformation. In this case, the civilized countries decide to take a new tour from 'economic environment' towards "environmental economics". The necessity of such change is empowerment and supports of specialists of natural resources and environmental issues along with economic experts governmental and policymakers. Nowadays, conservation of environmental resources is along important subjects to gain sustainability all around the world.

The economic valuation of natural resources is a perquisite tool which helps governmental planners and policymakers to take necessary actions for protecting environment. This approach would discover environmental goods, demand curves and also consumers' willingness to pay for environmental services which is of a big value (Bateman and Willis 1999; Haghjou et al. 2015; Haghjou et al. 2019).

Forests provide extended services to the human society, and population growth increased demands for these services; consequently, deforestation rates increased. According to the statistics, deforestation rates have a direct and straight relationship with the population growth rate (FAO 2010). This means every year a big share of forests benefits to human being and the universe is vanishing. Concerning this fact, awareness of forests services and their economic value would notice community about their loss during deforestation process and would increase their will and willingness to pay for protecting forest resources. In Iran, deforestation caused about 1 billion rial damages in 2002, which was about 0.8 per cent of country's GDP, based on World Bank's estimations (World Bank 2005).

Currently, about 4.5% of Iran is covered by forests. A time trend review of the forest cover shows that in the twentieth century about 18 million ha were covered by forests, which went down to 12 million ha and now only 7.3 million ha is covered (Rashidghalam 2019). Arasbaran Forests includes more than 87% of East Azerbaijan's forest area (they are about 164,000 hectares) and more than 54% of it is considered as protected area. Arasbaran Forests possess a rich biodiversity and they are the only habitat for the rare black rooster bird in the world. Due their important role as a habitat to the various species of animals and plants and woods, UNESCO declared Arasbaran Forests as one of the reserve of 10 "Biosphere" reserves in Iran since 1976. Based on Arasbaran's natural, historical, aesthetic and touristic values they attract many visitors every year and they have an important role in the sustainable development of the region (Department of Natural Resources in East Azerbaijan 2003).

The non-consumption functions are among numerous values which forests have; "refugium function" which means protection of animal species and providing a habitat for them is among non-consumption services. This function is one of the most important services which a forest can provide, since it is important in the aspect of tourism, natural balance and their own existent value which is important to the society. In this case, Arasbaran Forests are considered a natural treasure preserve valuable animal species which are so important to the society in the material and spiritual ways. Statistics show that due to damages and abnormal hunting, animal species of the Arasbaran Forests are in the danger of extinction (Department of Natural Resources in East Azerbaijan 2003). According to these facts, this research tries to estimate economic value and compensation surplus of three selected animal species of Arasbaran Forests. Since environmental conservation without public aid and financial supports—especially in the developing countries with fragile economic—is not possible, attaining consumers' willingness to pay for environmental services is a very important step in protecting environment including forests.

In this respect, many studies tried to research about this subject in recent years. Contingent Valuation Methods (CVM) is one of the non-market valuation methods for economic evaluation of environment and forests' services and due to its simplicity some studies applied this method (Sattout et al. 2007; Khodaverdizadeh et al. 2008; Barala et al. 2008; Pattison 2009; Molaei 2009; Jahanshahi and Mousavi 2011; Tao et al. 2012).

In some studies, the choice experiment (CE) method was applied for economic valuation of environmental good and services. Despite its complexity, this approach is becoming more common during recent years due to its advantages over CVM (Meyerhoff et al. 2009; Taylor and Longo 2010; Salehnia 2011; Wallmo and Lew 2011; Cerda et al. 2013; Haghjou et al. 2019). Some other researchers used contingent ranking approach for economic valuation of environmental resources. This approach, like the CE is complicated but more complete (Garrod and Willis 1997; Kumar and Kant 2007; Haghjou et al. 2015). Some researchers aimed the touristic and recreational function of the environment, and in this respect, they applied the travel cost method (Chae et al. 2012; Hayati et al. 2011). Finally, there are some researches who insisted on comparing valuation methods to see which one is more applicable and better (Sayadi et al. 2005; Bateman et al. 2006; Mogas et al. 2009).

In sum, it can be said that even though studies showed different effects for different demographic variables, but in total economic–demographic factors along with features of studied natural resource, and respondents' perspective towards environment could affect their WTP for different functions of environmental resource, including habitat function. Moreover, it could be added that environmental services, counting the refugium service, is significantly valuable for the society, and consumers are willing to pay a noteworthy amount to preserve it.

As it was mentioned, refugium function is one of the most valuable services of Arasbaran Forests, and gaining public financial supports is an important way for preservation of endangered animal species. As a result, this study concentrates on estimating respondents' WTP for selective animal species of Arasbaran Forests and their compensation surplus for protection and improvements in their applying Choice Experiment approach. In this study, among all the various animal species living in the forests, through some interviews with experts and conducting pre-tests, we selected tiger, black rooster and bear.

2 Materials and Methods

From recreational opportunities to direct inputs into the production process, environmental resources provide a complex set of values to individuals and benefits to society. The economic value is a human-oriented value and is considered as a valuable tool which is connected to the human utility maximizing. This means that it is designed based on people's desire and their preferences. If there was a direct market for environmental goods and services, using the normal pricing methods to value such services, would be possible. But the lack of a suitable market for many environmental functions makes usage of these methods almost impossible. Hence, one of the best methods of estimating non-market services is stated preference methods.

Stated preference approaches to nonmarket valuation rely on answers to carefully worded survey questions. These methods attempt to measure people's willingness to pay directly. The stated preferences approach relies on the data which are collected through direct questioning of respondents and their preferences. The methods consist of several valuation techniques. The common feature of all these techniques, are the direct questions from people about their possible choices, in a hypothetical market. This approach includes "contingent valuation methods" (CVM) and "Choice Modelling" such as: "choice experiments" and "contingent ranking". In this study, the CE method is applied. In this approach, the respondent is asked to choose his most preferred alternative among others. This method can provide much more information than methods like contingent valuation and every characteristic of the valuated resource, would have a value for itself. However, this method is more complicated compared to other ones. CE could result in welfare-adapted estimations, provided that the status quo option is one of the alternatives in the choice series, So that if the respondent is not interested in any of the improvement alternatives, he/she chooses the status que (Liu and Wirtz 2010; Haghjou et al. 2019).

CE has a common theoretical framework with other environmental valuation approaches using the random utility model (McFadden 1973). Based on this framework, the indirect utility function for each respondent i (U_{ij}) can be divided into two parts: (A) a systematic component (μ_{ij}), that is defined as a linear index of the attributes (X) of each j animal value in the choice set, and (B) a stochastic part (ε_{ij}), which stands for unobservable effects on individual's choice:

$$U_{ij} = \mu_{ij}(X_{ij)} + \varepsilon_{ij} = bX_{ij} + \varepsilon_{ij} \tag{1}$$

The respondent i would choose the animal species h other than animal species k, providing that $U_{ih} > U_{ik}$. Therefore, the odds of ranking h higher that k could be explained through Eq. (2):

$$\exp\{U_{ih} - U_{ik}\} \tag{2}$$

It is required to know the distribution of the error terms to derive an explicit expression for this probability. The typical assumption is that the error term ε_{ij} is supposed to be independent and identically distributed, and giving that:

$$\text{Prob}\big(\varepsilon_{ij} < t\big) = \exp\{-\exp(-t)\} \tag{3}$$

This distribution of the error term states that the probability of any particular alternative h being chosen as the most preferred can be expressed in terms of the logistic distribution (McFadden 1973) showed in Eq. 4 which is known as the conditional logit model:

$$P(U_{ih} \succ U_{ik}) = \frac{\exp(\lambda\mu_{ih})}{\sum_j \exp(\lambda\mu_{ij})} \tag{4}$$

which λ is a scale parameter, inversely proportional to the standard deviation of the error distribution.

This model can be estimated by conventional maximum likelihood procedures, with the respective log-likelihood functions stated in equation below, where y_{ij} *is* an indicator variable which takes a value of one if respondent j chose option i and 0 otherwise:

$$\log \text{L} = \sum_i \sum_j y_{ij} \log \frac{\exp(\lambda\mu_{ih})}{\sum_j \exp(\lambda\mu_{ij})} \tag{5}$$

After estimation of parameter the willingness to pay of individuals or the implicit price of each attribute is derived through the final rate of substitution between non-monetary and monetary attributes and it is estimated from the ratio of non-monetary factor to the monetary one (Hanley et al. 2002), which is shown as:

$$\text{Marginal WTP} = -\left[\frac{\beta\ \text{non-monetary}}{\beta\ \text{monetary}}\right] \tag{6}$$

Moreover, the compensation variation or the compensation surplus (w) which actually is the difference between before and after the change in utility of the individual because of betterment in the animal species' condition could be shown in Eq. 7. The coefficient μ gives the marginal utility of income and is the coefficient of the cost attribute (Hanley et al. 2002; Liu and Wirtz 2010; Haghjou et al. 2015; Haghjou et al. 2019):

$$w = \frac{1}{\mu}\left[\ln \sum_{i \in C} e^{V_{i0}} - \ln \sum_{i \in C} e^{V_{i1}}\right] \tag{7}$$

It should be noted that an important assumption of this specification is that the selections from the choice set must follow the "independence from irrelevant alternatives" (IIA) property. This states that the relative probabilities of two options being selected are unaffected by the introduction or removal of other alternatives (Hanley et al. 2002). This condition could be tested through the test introduced by Hausman and McFadden (1984) in which the model is estimated in unrestricted way by all options, and then one alternative is removed from the model and it would be estimated again (restricted model; finally, the Hausman statistic is calculated via the definition 8 brought below, which follows the χ^2 distribution and its significance could be confirmed via the calculated χ^2 table:

$$H = (\widehat{\beta}_r - \hat{\beta})'(\hat{v}_r - \hat{v})^{-1}(\widehat{\beta}_r - \hat{\beta}) \overset{H0}{\sim} \chi^2(m) \tag{8}$$

The first and most important step in the CE method is designing of choice cards. To this end, the main attributes of the resource and the level of each attribute is identified, then the cards and, henceforth, the questionnaire are designed according to the characteristics of the test. In practice, attributes are selected from reviewing of previous studies or interviewing with experts group (target group). It should be noted that the price paid for the studied resources are one of the reviewed attributes and, through the monetary factor, it is possible to estimate the willingness to pay for each attributes of the forest. Also the levels of each attribute are identified through the exploratory studies, literature reviews and interviews with the target groups.

The statistical design theory is used for level composition and formation of appropriate scenarios to present to the respondents. Complete factorial design is one of the available options in this stage; however, because of a large number of compounds in this technique, alternative methods such as "Partial factorial design" is used in which the number of possible combinations are greatly reduced. Table 1 shows the selected attributes (animals) in valuation of Arasbaran Forests' animal species. As mentioned before, the attributes were chosen through interviewing with environmental experts and literature review. The refugium functions of Arasbaran Forests are

Table 1 The attributes of Arasbaran forests' selected animal species and their levels

Bid price (Rial)[a]	Three protected animal species: bear, black rooster, tiger	Attributes
5000	Critical condition	Levels
15,000	Relative improvement	
25,000		
35,000	Optimum condition	
45,000		

[a]During this study, one Rial was about 3.2×10^{-5} Dollar

divided into three attributes (three main animal species) with three levels (crisis, relative improvement and optimum), while the bid price has five selected levels (5,000, 15,000, 25,000, 35,000 and 45,000 Rial)[1].

In this regard, to design the questionnaire, 12 alternatives and 6 choice set was determined which were collected in 2 trio-blocks. We used SAS 9.2 software for designing the cards. Each choice set, includes the relative improvement and optimum conditions of animals besides one status quo option.

Each level has specific traits which are shown in the Fig. 1. These were chosen through interviewing with target group (environmental experts) and studying the conservation plans of Arasbaran Forests. These characteristics were presented to the respondents before they answer to the questionnaire.

After noticing respondents about the characteristic of animal species, selection cards were presented to the respondents. Figure 2 shows an example for the cards. Through these cards, the willingness to pay for the three selected animal species of Arasbaran Forests and their levels could be estimated.

The questionnaire was conducted among 334 respondents which were randomly selected among visitors of the Arasbaran Forests along with some individuals from ten neighbouring cities (which all were within a radius of 250 km from the forests). The cities all were from three adjacent provinces: West Azerbaijan, East Azerbaijan and Ardabil. It should be noted that the sample size is calculated using the formula introduced by Orme (1998):

Attributes ⇩ / Levels ⇨	Optimum condition	relative improvement	Critical condition
(Selected animals: Tiger, Black Rooster, Bear)   	Protecting and improvement of reserves; increasing reproductive powers of all animal species specifically the variable ones, construction of some places for some animals which are not capable enough to regenerate in normal ways; establishing some exclusive centres for continuous health studies of animals	Relative improvement some conditions of forest's ecosystems and improving living conditions of animal species, creation of some temporary moving clinics for animal species	Status Quo **(present condition)** animal species' lives are in danger and there is not any proper management and planning for their protection

Fig. 1 The characteristics and their levels of selected animal species of Arasbaran forests

[1]It should be noted that the bid prices are selected through a pre-test.

⇩ Attributes /Alternatives ⇨	Alternative C	Alternative B	Alternative A
 The Black Rooster	I do not want any change in the current situation and I am not willing to pay any fee for it.	Relative Improvement	Optimum condition
 The Bear		Optimum	Critical condition
 The Tiger		Optimum	Critical condition
WTP (Rial)		25000	5000
Which alternative would you like to choose? Please mark it.	•	•	•

Fig. 2 An example of the selection cards to estimate the value three selected animal species' of Arasbaran forests

$$N = 500(\frac{N_{le}}{N_{alt}N_{rep}}) = 500(\frac{3}{3.3}) \approx 167$$

in which N_{lev} is the highest number of levels per attribute, N_{alt} is the alternatives number in each choice set and N_{rep} is the number of questions that each respondent should answer (Orme 1998).

3 Results and Discussions

Table 2 shows descriptive statistics of respondents' characteristics. According to this table, respondents mostly are middle-aged, and married men who belong to the families with small divisions and their average annual visits from the forests are less than once a year. Their gross annual income is a middle amount. The index of friendly attitude towards Arasbaran Forests which measures respondents' perspective about these forests is an criteria that consisted of 10 speeches to measure respondents' point of view towards Arasbaran (e.g. "I am willing to waiver some of my welfare to protect Arasbaran forests"); to evaluate these speeches, some codes from 5 (very important) to 1 (not important) were used. The average of this index highlights the importance of Arasbaran Forests to the respondents. The individuals' level of education measures through an ordinal variable which is ranked through some sequential codes from:

Table 2 Descriptive statistics of explanatory variables

Variable	Mean	SD	Min	Max
Income of respondents (10,000 Rial)	1691	740.761	250	6000
Age of respondents	40.40	7.700	23	71
Gender of respondents	0.729	0.445	0	1
Respondents' level of education	5.858	1.010	4	8
Family dimension	3.456	1.300	1	7
Visits per year	0.631	0.730	0	3
Respondents' friendly attitude towards Arasbaran forests	3.714	0.680	2	5

1 = Illiterate, to 8 = Ph.D. This variable's average indicates an academic level of education for most of respondents.

The estimation results of conditional logit regression model are shown in the Table 3. The model results will be applied to estimate the value and compensation

Table 3 Standard and synthetic estimation conditional logit model for three animal species of Arasbaran forests

Variable	Standard conditional logit regression		Conditional logit regression with interactions (synthetic model)	
	Coefficient	Standard error	Coefficient	Standard error
ASC	1.730***	0.251	1.690***	0.000
price	−0.00,022***	0.000	−0.001***	0.036
Black rooster's optimum condition	0.387***	0.045	0.404***	0.092
Black rooster's relative Improvement	0.306***	0.099	0.210***	0.071
Bear's optimum condition	0.250***	0.083	0. 124***	0.101
Bear's relative improvement	0.230***	0.086	0.293***	0.104
Tiger's optimum condition	0.450***	0.097	0.529***	0.121
Tiger's relative improvement	0.323***	0.114	0.398***	0.000
Price × education level	–	–	0.00002***	1.41e−08
Price × income	–	–	3.87e−08***	0.0000224
Price × firendly attitude towards Arasbaran	–	–	0.0001***	0.0000189
P × Annual visits per year	–	–	0.00007**	0.262
	LR chi2 = 459.34, Pseudo-R^2 = 0.32		LR chi2 = 650. 86, Pseudo-R^2 = 0.37	

***: $P < 0.01$, **: $P < 0.05$, *: $P < 0.1$

surplus of selected animals and their levels. In the table, ASC[2] shows alternative special constant and price represents the bid price.

Concerning that all economic–demographic variables are fixed and unchanged for all answers of a specific respondent, to infer the effects of these personal factors on his/her WTP, we used the synthetic model estimation which uses interactions of bid price (p) and other variables (income, education of respondents, friendly attitude towards forests and the number of annual visits) to make variation for comparing and analysing. Table 3 shows the estimation results of both models.

It should be added that the Hausman–McFadden test for IIA assumption is considered for this estimation, approved the independency of irrelevant alternatives in the model. Hence, the results of conditional logit model could be interpreted.

As shown the Table 3, all levels of animal species have positive sign, while the bid price, as expected, has significant and negative sign. The coefficients of optimum levels of each animal species have higher value and consequently higher utility than relative improvement levels. Chi-square test[3] between standard model and interaction model confirms that the model with interactions is better than the other one.

Results from the alternative special constant (ASC) in both models indicate that regardless of the type and the amount, in terms of a protection, respondents are willing to pay to protect the animal species' condition and such preserving programmes could significantly increase their utility. Moreover, based on the results of synthetic model's estimation, individual's educational level, his/her gross income, respondent's number of visits per year from the Arasbaran besides the individual's friendly perspective about the forests have a direct and positive relationship with respondent's willingness to pay for the protection of animal species' conditions. In this respect, there are other studies which reached the same results about these factors (Bateman et al. 2006; Sattout et al. 2007; Mogas et al. 2009; Tao et al. 2012; Haghjou et al. 2015; Haghjou et al. 2019).

Through the estimation results of synthetic model which was proven to be the better one, and using the equation along with consideration of studied area's population, the total (for whole population), annual and monthly willingness to pay for the chosen animals and their level of protection are estimated (Table 4). According to this table, respondents declared black rooster and tiger as the least and the most valued animals, respectively. Results reveals that, the total annual value of these three chosen animal species is around 692.655×10^9 Rial (About 22×10^6 USD), in which around 50% of it belongs to tiger, and about 29 and 21% of this value is attached to the bear and blackcock, respectively. This could be because of informing and advertising programs about tiger's (even in some cases the bear's) value and its being endangered in the mass media and press.

Finally, Table 5, shows the compensation surplus (CS) which are calculated using Eq. (7). Based on the results, improvement in situation of three chosen animal species

[2]ASC indicates choosing of each improvement alternative, against the current situation or the status quo. It gets the value of one, in case any of preserving programmes are chosen by the respondent and gets the zero if not so.

[3]$LR_{test} = -2 \times (log - likelihoodfunction_{unr} - log - likelihoodfunction_{res}) \sim \chi^2$

Table 4 Calculation results of willingness to pay for three selected animal species and their levels

Animals and their levels	Tiger's optimum condition	Tiger's relative improvement	Bear's optimum condition	Bear's relative improvement	Black rooster's optimum condition	Black rooster's relative improvement
Ind. monthly WTP (Rial)	3520	2650	2370	1820	1720	1400
Ind. annual WTP (Rial)	42,240	31,800	28,440	21,840	20,640	16,800
Total WTP (Rial)	422.40×10^9	256.10×10^9	229.04×10^9	175.89×10^9	166.22×10^9	135.30×10^9
Mean of two levels (Rial)	339.25×10^9 (11×10^6 USD)		202.645×10^9 (6.3×10^6 USD)		150.76×10^9 (4.7×10^6 USD)	
Ranking of features	1		2		3	

Table 5 Results of SC extracting of relative and optimum improvement in three chosen animals' condition

Optimum condition of tiger	Relative improvement of tiger	Optimum condition of bear	Relative improvement of bear	Optimum condition of black rooster	Relative improvement of black rooster
2.3×10^6 Rial (0.71×10^3 USD)	1.8×10^6 Rial (0.57×10^3 USD)	1.5×10^6 Rial (0.41×10^3 USD)	1.2×10^6 Rial (0.32×10^3 USD)	1.1×10^6 Rial (3.4×10^2 USD)	9.3×10^5 Rial (2.9×10^2 USD)
5.07×10^7 Rial (1.6×10^3 USD)	Compensation surplus of optimum improvement in the three animal species' condition		3.9×10^7 Rial (1.1×10^3 USD)	Compensation surplus of relative improvement in the three animal species' condition	

could upgrade the SC of the respondents for about 3.9×10^7 Rial (about 1.1×10^3 USD), which could be increased up to 5.07×10^7 Rial (about 1.6×10^3 USD) in the case the improvements in the all animal's situation reaches to the optimum condition.

4 Conclusion

The key to sustainable development is the conservation of valuable natural resources, including forests and their valuable treasures. Achieving the goals and sustainable management of conservation, rehabilitation and development in the natural resource, especially forests and their valuable treasures like animal species, according to many scholars, could be possible through educating people, organizing and meeting expectations and, most importantly, attracting people's participation in natural resources and strengthening the working spirit of the them. This study tries to meet this goal through evaluation of people's WTP for conservation of tiger, bear and black rooster, which are considered precious functions of Arasbaran Forests' resources. Based on the results, public willingness to pay and also their compensation surplus for three animals of the Forests (tiger, bear and black rooster) are large and notable amount (About 22×10^6 USD and about 1.6×10^3 USD, respectively). According to this fact, the refugium value and conservation of animal species of Arasbaran Forests are of a great value for the society.

In this research, only three animal species were economically valuated, while Arasbaran Forests provide habitat for a large number of animals; yet the evaluation price was a big and considerable amount. This gives a hint to the policymakers for animal species conservation planning and attract public aid. Moreover, prioritizing and ranking of studied animals indicates that the tiger and the bear are more favourable, respectively. Such results might be due to effects of mass media and social programmes in introducing and informing about these animal species. This shows the importance of advertising programmes to protect the environment and animal habitats. Also, this result can be useful in plan-making for the priory of their conservation programs. Moreover, creating a natural zoo in the Arasbaran Forests, will increase visitors information about the various animal species into the area, but it can also be a financial aid to the protection and development plans.

Concerning that the respondents' level of income significantly increased their willingness to pay for the protection sceneries, any effort and plan for increasing the financial power of the region's society, would increase the financial aid for protection of the forests and animal species. Finally, considering the significant and positive relationship between respondents' friendly attitude towards animal species and their WTP, any programme such as relative NGO's and supporting their activities, or making informative TV programmes or billboards to augment public awareness in this respect, could be a useful step for protection of animal species and all environmental treasures.

References

Barala NM, Sternb J, Ranju B (2008) Contingent valuation of ecotourism in Annapurna conservation area, Nepal: implications for sustainable park finance and local development. Ecol Econ 66:218–227

Bateman IJ, Willis KG (1999) Valuing environmental preferences: theory and practice of the contingent valuation method in the US, EU, and developing countries. Oxford University Press, Oxford

Bateman IJ, Cole MA, Georgiou S, Hadley DJ (2006) Comparing contingent valuation and contingent ranking: a case study considering the benefits of urban river water quality improvements. J Environ Manage 79(3):221–231

Cerda C, Ponce A, Zappi M (2013) Using choice experiments to understand public demand for the conservation of nature: a case study in a protected area of Chile. J Nat Conser 21(3):143–153

Chae D, Wattage P, Pascoe S (2012) Recreational benefits from a marine protected area: a travel cost analysis of Lundy. Tour Manag 33(4):971–977

Department of Natural Resources in East Azerbaijan (2003) The preservation plan of Northern Arasbaran forests (in Farsi)

FAO (2010) State of the world's forests. Food and Agriculture Organization, Rome

Garrod GD, Willis KG (1997) The non-use benefits of enhancing forest biodiversity: a contingent ranking study. Ecol Econ 21(1):45–61

Haghjou M, Hayati B, Pishbahar E, Molaei M (2015) Using the contingent ranking approach to assess the total economic valuation the of Arasbaran forests in Iran. Tai Jol For Res 31(2):87–102

Haghjou M, Hayati B, Pishbahar E, Molaei M (2019) Estimating the non-use values and related compensative surplus of Arasbaran forests in Iran: an application of the choice experiment method. In: Rashidghalam M (ed) Sustainable agriculture and agribusiness in Iran. Perspectives on development in the Middle East and North Africa (MENA) Region. Springer, Singapore

Hanley N, Wright R, Koop G (2002) Modelling recreation demand using choice experiments: rock climbing in Scotland. Environ Resour Econ 22:449–466

Hayati B, Salehnia M, Hoseinzadeh J, Dashti G (2011) Estimating the recreation value of Fadak Park of Khoy City: an application of individual travel cost method. In: The first conference of the Iranian urban economy, Mashhad, 23–24 Nov (in Farsi)

Hausman J, McFadden D (1984) Specification tests for the multinomial logit model. Econometrica 52(5):1219–1240

Jahanshahi D, Mousavi N (2011) The economic valuation of environmental amenities, case study: Yasouj waterfall. In: The first international conference on tourism management and sustainable development, Marvdasht, 29–30 Sep (in Farsi)

Khodaverdizadeh M, Hayati B, Kavousi M (2008) Estimating the outdoor recreation value of Kandovan tourism village of East Azarbayjan with the use of contingent valuation method. J Environ Sci 4:43–54 (in Farsi)

Kumar S, Kant S (2007) Exploded logit modeling of stakeholders' preferences for multiple forest values. For Pol Econ 9(5):516–526

Liu X, Wirtz KW (2010) Managing coastal area resources by stated choice experiments. Estuar Coast Shelf Sci 86:512–517

McFadden D (1973) Conditional logit analysis of qualitative choice behaviour. In: Zarembka P (ed) Frontiers in econometrics. Academic Press, New York

Meyerhoff J, Liebe U, Hartje V (2009) Benefits of biodiversity enhancement due to nature-oriented silviculture: evidence from two choice experiments in Germany. J For Econ 15(1–2):37–58

Mogas J, Riera P, Bennett J (2009) A comparison of contingent valuation and choice modeling with second-order interactions. J For Econ 12(1):5–30

Molaei M (2009) Ecological economic valuation of Arasbaran forest. A PhD thesis, Department of Agricultural Economics and Development, University of Tehran (in Farsi)

Orme B (1998) Sample size issues for conjoint analysis studies. Sawtooth Software technical paper. www.sawtoothsoftware.com

Rashidghalam M (2019) Introduction to sustainable agriculture and agribusiness in Iran. In: Rashidghalam M (ed) Sustainable agriculture and agribusiness in Iran. Perspectives on development in the Middle East and North Africa (MENA) Region. Springer, Singapore
Salehnia M (2011) Estimating willingness to pay for improvement in Lake Urmia's environmental situation using choice experiment. An Msc Thesis, Faculty of Agriculture, Department of Agricultural Economics, Tabriz University (in Farsi)
Sattout EJ, Talhouk SN, Caligari PDS (2007) Economic value of cedar relics in Lebanon: an application of contingent valuation method for conservation. Ecol Econ 61:315–322
Sayadi S, Roa CG, Requena JC (2005) Ranking versus scale rating in conjoint analysis: evaluating landscapes in mountainous regions in south eastern Spain. Ecol Econ 55(4):539–550
Pattison JK (2009) The non-market valuation of wetland restoration and retention in Manitoba. MSc thesis in Agricultural and Environmental Economics, University of Alberta, Canada
Tao Z, Yan H, Zhan J (2012) Economic valuation of forest ecosystem services in Heshui Watershed using contingent valuation method. Procedia Environ Sci 13(2012):2445–2450
Taylor T, Longo A (2010) Valuing algal bloom in the Black Sea Coast of Bulgaria: a choice experiments approach. J Environ Manag 91(10):1963–1971
Wallmo K, Lew D (2011) Valuing improvements to threatened and endangered marine species: an application of stated preference choice experiments. J Environ Manage 92(7):1793–1801
World Bank (2005) Islamic Republic of Iran cost assessment of environmental degradation. Report No. 32043-IR

Maryam Haghjou is a lecturer in Department of Agricultural Economics at University of Tabriz. She holds her B.S., M.Sc. and Ph.D. from Department of Agricultural Economics, University of Tabriz. She has taught a number of courses on Natural Resource Economics and Microeconomics. Her research and publication has focused on Sustainable Development, Environmental Management and Natural Resource Management. She has publications in Journal of Agricultural Science and Technology (JAST).

Babollah Hayati is a Professor in Department of Agricultural Economics at University of Tabriz. He was dean of Faculty of Agriculture during 2015–2018. He holds a B.Sc. in Agricultural Economics at University of Tehran and a M.Sc. and Ph.D. in Natural Resource Economics at Tarbiat Modares University. His areas of special interest are Natural Resource Economics, Sustainable Development Economics and Microeconomics. His recent publications have appeared in numerous journals including the Journal of Agricultural Science and Technology (JAST) and Engineering Sustainability.

Esmaeil Pishbahar is Associate Professor of Agricultural Economics at University of Tabriz, Iran. He holds a B.Sc. in Agricultural Economics from University of Tabriz and a M.Sc. in Agricultural Economics from University of Tehran. He did his Ph.D. in Science Economics at departments of Economics and Management, University of Rennes 1, France. His areas of interest and research are Applied Econometrics, Agricultural Risk Management and Insurance, and International Trade. His teaching area are Advanced Econometrics, Mathematical Economics, and Macroeconomics at under- and postgraduate levels. He has over 100 publications in journals and chapters in books.

Morteza Molaei is an Associate Professor in Department of Agricultural Economics at Urmia University. He is one of the pioneers of Department of Agricultural Economics at Urmia University. He was head of department during 2013–2017. He received his B.Sc. degree in Agricultural Economics from University of Tabriz; M.Sc. and Ph.D. degrees from University of Tehran. Dr. Molaei's areas of expertise are Environmental Impact Assessment, Sustainable Development,

Environmental Efficiency and Water Resource Economics. He teaches Econometrics and Environmental Economics at postgraduate level. Currently he works on a FAO project "Farm and Household Livelihood Survey Covering Selected Sub-basins within Lake Urmia Basin".

Factors Affecting Consumers' Awareness of Pesticides-Free Fruits and Vegetables

Maryam Haghjou, Babollah Hayati, and Esmaeil Pishbahar

Abstract Chemical pesticides are among the most important barriers to sustainability in agriculture. Their numerous disadvantages for the environment and living organisms, as well as the food cycle, pushed developed societies to eliminate or reduce pesticides application. Meanwhile, consumers, due to the health problems, have changed their approach to safe and pesticide-free food products. However, in developing countries such as Iran, the lack of knowledge and awareness of safer food products is one of the main reasons for the lack of organized market for such products. In this study, which was conducted in 2010 among 394 consumers from Marand City (Iran), we tried to examine consumers' awareness of pesticide-free fruits and vegetables and its determinants using ordered probit model. Data were collected through the field study and the questionnaire. According to the results, only 20% of the respondents have appropriate information about the features of pesticide-free fruits and vegetables, and about 24% have low information or lack of awareness. Estimation results show that factors such as educational level, positive environmental tendencies and adherence to healthy lifestyle index among individuals, as well as having children under the age of ten or people with specific diseases in the household have a positive and significant impact on awareness of pesticide-free fruits and vegetables. In this context, female respondents' awareness was more than males. In this respect, appropriate advertising, conducting training courses for all levels of education, raising community awareness of sustainability issues and safer products besides environmental and a healthy lifestyle issues are suggested.

Keywords Awareness · Effecting factors · Marad city · Ordered probit · Pesticide-free fruits and vegetables

---

M. Haghjou (✉) · B. Hayati · E. Pishbahar
Department of Agricultural Economics, University of Tabriz, Tabriz, Iran
e-mail: maryam.haghjou@yahoo.com

B. Hayati
e-mail: b-hayati@tabrizu.ac.ir

E. Pishbahar
e-mail: pishbahar@yahoo.com

M. Rashidghalam (ed.), *The Economics of Agriculture and Natural Resources*,
Perspectives on Development in the Middle East and North Africa (MENA) Region,
https://doi.org/10.1007/978-981-15-5250-2_9

1 Introduction

Sustainable agriculture system means cultivating according to characteristics of ecosystem services and considering the relationships between organisms and their physical environment. It has been defined as "an integrated system of plant and animal production practices having a site-specific application that will last over the long term." Besides satisfying of human food and economic sustainability of agricultural practices, sustainable agriculture concerns about enhancement of environmental quality and the natural resource based on the agricultural economy's principles, as well as increasing the quality of life for society as a whole (Gold 2018). According to scientific findings, pesticides are the most harmful chemical inputs in food products and their most harmful effects occur in fruit and vegetable products. The extreme damages, which these artifacts bring to the environment and natural ecosystems along with undeniable effects on consumers' health, make the pesticide as one of the biggest barriers to sustainable agriculture and human and environmental health. This causes high concern among policymakers and consumers about the excessive use of these materials in agricultural production (Howard and Berry 2008; Haghjou et al. 2013).

Regarding the negative impacts of pesticides on human health and the environment, their use in developed countries is significantly different from developing countries. These countries have reduced their use of pesticides by applying other pest control methods such as biological control and organic farming, but such approaches still have not been expanded in developing countries (Erfanmanesh and Afyouni 2007).

Statistics on pesticide's use shows that more than 4,200,000 tons of active ingredients are used worldwide in 2016. China consumes about 1,700,000 tons followed by the United States (407,000 tons). Currently, about 30,000 tons of pesticides are used annually in Iran. Low consumption of pesticides in the country is directly related to the low level of agricultural production.[1] Despite the national action plan to reduce the use of pesticides in the country, the amount of applying these chemical substances in agricultural production is still high and is at dangerous level (Erfanmanesh and Afyouni 2007). According to Iranian Statistic Center (2012), the use of pesticide in East Azerbaijan province was increased by 16% in 2006.

The lack of farmers' knowledge on pesticides and their safe use in terms of amount and time (taking into account the time of the worms of toxins) has made consumers uncomfortable with the taste of fruits and vegetables. Although the majority of consumers are concerned about the unfavorable taste of crops, public awareness of the cumulative effects of absorbed toxins in the body and its harmful and harmful effects, which leads to dangerous diseases, is at low level. Limiting the use of pesticides and chemical fertilizers is essential to move toward sustainable agriculture. Therefore, it is necessary to educate the farmers and consumers, and to make them aware of health risks caused by the use of pesticides (Mahmoudi and Mahdavi-Damghani 2010).

[1] http://agricultech.com/node/154.

The eagerness of consuming healthy food has grown steadily due to the increasing concerns about contamination of conventional products besides environmental issues, and the export market for such products has become widespread among affluent societies. However, in Iran, there is no tangible move toward planning, leading, and supporting the production of safer and pesticide-free foods. This will deprive farmers of the benefits of exports and higher income, and also the entire society, from improving their nutrition, health, and environment. This is despite the fact that in our country due to the dry climate conditions and the abundance of labor force, the production of healthier products seems more economic and easier than in many other parts of the world (Mahmoudi and Mahdavi-Damghani 2010).

All of these issues suggest that most people every day, based on their dietary patterns, consume significant amounts of pesticide residues, which in the long run cause many health problems. However, the production and supply of pesticide-free products is still not fully implemented anywhere in the country and even at the provincial level, and is still in the preliminary stage. The first step in addressing and moving toward the production of such products is the identification of the target, i.e., consumers. The reason is that the main drivers of economic growth is the management knowledge based on the understanding of consumers' behavior; the consumers and their knowledge of the product that he/she buys, is always a developer and prosperous of the market. Hence, this study is concerned about the awareness of consumers about pesticide-free fruits and vegetables (PFFV) and its determinants.

The importance of consuming healthy foods has led to many studies in recent years to examine consumers' attitudes and tendencies. The results of studies by Chinnici et al. (2002), Harper and Makatouni (2002) and Boxall et al. (2007) showed that consumers have a favorable opinion on this type of products and one of the most important reasons is the lack of health risks and the safety of these products comparing the conventional ones. Sachs et al. (1987) and Yiridoe et al. (2005), and Haas et al. (2013) assess the consumer awareness of the harmful chemicals and pesticides effects as well as the importance of health criteria as factors influencing their attitude and willingness to pay for safer food products. Studies of Yiridoe et al. (2005), Rodriguez et al. (2007) and Boxall et al. (2007) found that factors such as better taste and naturalness have positive effects on willingness of people to buy healthy foods. While others such as Millock et al. (2005) and Cowan et al. (2000), have shown that factors such as experience of consumption and direct comparison of these products with conventional ones as well as advertising are influential factors. Another group of studies by Torjusen et al. (2010), Darby et al. (2006), and Ahmad and Juhdi (2008) have indicated that female, young consumers, people with a job or educational history in agriculture, and also people with a high protective attitude toward environment have positive visions. Therefore, they tend to buy organic and safer food at higher prices. According to Shafie and Rennie (2012), nutritional values, the importance of individual health, environmental inclinations, along with individual–social characteristics of consumers such as age, gender, and educational level, can be viewed as effective factors on their perspective on safer and organic food products. Another study by Tison (2012) considers appropriate and informative labels to be a factor in the positive tendency of individuals toward organic food products. In Iran, some studies carried out to investigate the type of attitude, level of knowledge, and willingness to pay for healthy food (e.g., Ghorbani et al. 2007; Alizade et al. 2008; Haghjou et al. 2013; Hayati et al. 2017).

According to the literature review, studying consumers' attitude and awareness, about safer products and factors affecting it, is an effective step to improve community's information about them and developing of markets for healthy food products in the country. In this regard, this study investigates the consumers' awareness of pesticide-free fruits and vegetables, and its determinants in Marand City of East Azerbaijan Province (Iran). This city is an important region in producing fruits and vegetables of the country.

The remainder of the paper is structured as follows: the next section presents the methodology used in the empirical analysis. The third section describes the data collection and presents the study design. The fourth section presents the results of the study. The paper concludes at fifth section.

2 Methodology

2.1 *Ordered Probit Model*

In many economic and social studies, there are situations where choices, and indeed dependent variables, are ranked or sequential. The specificity of these types of variables is that they are ordered and ranked and displaying different states with numbers is not just a separation in choice; i.e., a higher code or number in choosing an option, means respondent's higher preference. In such conditions, ordered models are applied to explain the ranked dependent variable's change (Pishbahar 2018).

Since this study's dependent variable also is an ordered one, we used these models to interpret its variations. These types of model are often modeled using unobserved or latent variables. To simplify, it is assumed that the observed responses through a latent variable Y^* are linearly correlated with the explanatory variable X as in Eq. 1:

$$Y^* = \beta_1 + \beta_2 X + u_i \tag{1}$$

This latent variable can be a person's "attitude" about a particular issue in the "strongly disagree" to "strongly agree" range. In the ordered models assuming that the values that the dependent variable takes are sequential, expanding Y^* into a ordered polynomial variable, taking into account the fact that it is a continuous variable, can simply be done as follows:

$$y_i = \begin{cases} 0\, if\ y^* \leq \mu_1 \\ 1\, if\ \mu_1 \prec y^* \leq \mu_2 \\ 2\, if\ \mu_2 \prec y^* \leq \mu_3 \\ \vdots \\ J\ if\ \mu_{j-1} \leq y^* \end{cases} \tag{2}$$

In which $\mu_{1\ \text{to}}\ \mu_j$ are the thresholds or unspecified limits which re-specified by the index variable d_{ij}.so that if Y_i belongs to alternative J, d_{ij} is 1 and 0 otherwise. To include the orders, and because all probabilities are positive, the following relationship must be established as:

$$0 \prec \mu_1 \prec \mu_2 \prec \cdots \prec \mu_{j-1} \tag{3}$$

Using Eq. (3) and normalization of $u_i(0, 1)$, probabilities are (Greene 2005; Pishbahar 2018):

$$\begin{aligned} &prob(y = 0|x) = F(-x'\beta) \\ &prob(y = 1|x) = F(\mu_1 - x'\beta) - F(-x'\beta) \\ &prob(y = 2|x) = F(\mu_2 - x'\beta) - F(\mu_1 - x'\beta) \\ &\vdots \\ &prob(y = j|x) = 1 - F(\mu_{j-1} - x'\beta) \end{aligned} \tag{4}$$

in which $F(.)$ shows Cumulative distribution function of u_i. In this type of models, the marginal effects are calculated to evaluate the effect of independent variables on the predicted probability of the dependent variable or to select the order of alternatives. Using the following equations, the marginal effects are calculated as:

$$\begin{aligned} &\frac{\partial prob(y = 0|\text{x})}{\partial x_i} = -F(-x'\beta)\beta \\ &\frac{\partial prob(y = 1|\text{x})}{\partial x_i} = \left[F(-x'\beta) - F(\mu_1 - x'\beta)\right]\beta \\ &\vdots \\ &\frac{\partial prob(y = j|\text{x})}{\partial x_i} = F(\mu_{j-1} - x'\beta) \end{aligned} \tag{5}$$

One of the assumptions of ordered is that the relationships at each level of the dependent variable are similar. In other words, by default the relationships that vary from low to high levels of dependent variable are the same and have the same slope. Basically, the ordered probit model estimates a similar equation on all the levels of the dependent variable and their deference is their intercept. In this model, we have

Table 1 Variable definition

AWR	Respondent's knowledge of PFFV index (Likert scales of 8 statements): 1 = Strongly disagree, 2 = Disagree, 3 = indifferent, 4 = Agree, 5 = Strongly agree
INC	Respondent's monthly income (RLs)
Age	Respondent's age (years)
FDIM	Respondent's family size (person)
ENV	Respondent's friendly attitude for environment index (Likert scales of 8 statements): 1 = Strongly disagree, 2 = Disagree, 3 = indifferent, 4 = Agree, 5 = Strongly agree
SShop	An index showing consumer's safer shopping criteria in buying fruit and vegetables (Likert scales of 5 statements): 1 = Strongly disagree, 2 = Disagree, 3 = indifferent, 4 = Agree, 5 = Strongly agree
EDU	Education of respondent (ordinal variable): 1 = Illiterate, 2 = Primary School, 3 = Junior high school, 4 = Senior high school, 5 = Associated Diploma (AD), 6 = B.Sc., 7 = M.Sc., 8 = Ph.D.
Gender	Gender of respondent (dummy variable): 1 = Male, 0 = Female
MATRI	Matrimony of respondent (dummy variable): 1 = Married, 0 = Single
SPCL	If respondent has children younger than 10 years old or People with special disease in the household (Dummy variable): 1 = Existence, 0 = Nonexistence

parallel regression lines for different levels, and is called "parallel lines assumption." If this assumption is violated, the ordered probit is not a suitable model to fit the data. To test the establishment of parallel regression lines, we use Brant test (Pishbahar 2018).

To depict factors influencing consumer's awareness about pesticide-free fruits and vegetables, we used the following model:

$$\begin{aligned} AWR_i = \beta_0 + \beta_1 \cdot INC_i + \beta_2 \cdot AGE_i + \beta_3 \cdot FDIM_i + \beta_4 \cdot ENV_i + \beta_5 \cdot SSHOP_i \\ + \beta_6 \cdot MATRI_i + \beta_7 \cdot Gender + \beta_8 \cdot SPCL + U_i \end{aligned} \tag{6}$$

Table 1 shows definition of variables in the estimated ordered probit model. In this study, we used Limdep 7.0 econometric software to estimate the ordered probit model.

2.2 Data and Survey Design

To develop study questionnaire, a small pre-test was taken from 50 respondents. Then based on the collected data, main survey was designed. The statistical population includes the households of Marand City. According to the latest official statistics, the city had about 68,313 households (Iranian Statistics Center 2012). We use the

Cochran formula to obtain the sample size. Finally, the questionnaire, through random sampling method and with in-person interviews, was conducted among 394 families of Marand City in 2010. The purpose of the survey was to collect data on respondents' awareness of pesticide-free fruits and vegetables along with some explanatory variables, such as demographic and attitudinal information of the respondents and their families.

In the first part of the questionnaire, some information was collected on respondents' attitudes. The first question tried to calculate individual's awareness and knowledge of pesticide-free fruits and vegetables (which we call it: knowledge of PFFV Index). To estimate this index, we applied an 8-point Likert scale (see Table 2). We asked respondents to evaluate each statement using a 5-level Likert item (5 = strongly agree; 1 = strongly disagree).

The second question tries to get consumers degree of concerning about safety factors in shopping and their knowledge of safe and healthy fruits and vegetables (Index of safe shopping). This question also was designed as Likert scales scoring from 5(= strongly agree) to 1(= strongly disagree), which includes five statements to evaluate individual's attention safe-shopping criteria (Table 3).

The third question evaluates respondents' friendly attitude toward environment (friendly environmental attitude index). This index also was a Likert scale consisted of eight statements which measure individuals' friendly attitude concerning environmental matters using 5(= strongly agree) to 1(= strongly disagree) scoring (Table 4). It should be mention that some statements had opposed meaning to encounter accidental answers by consumers. As it is clear, such statements had reversed scoring (1 = strongly agree; 5 = strongly disagree).

Table 2 Statements of respondent's knowledge of PFFV index

1	In the process of producing these products, chemical pesticides are not used
2	These products are healthier than conventional ones
3	In the production of these products, natural methods (such as the use of natural enemies of pests and mechanical methods) are used to deal with pests
4	Pesticide-free fruits and vegetables reduce the risk of diseases such as cancer
5	Due to the lack of chemical pesticides, these products have better flavor
6	Pesticide-free fruits and vegetables are more expensive than conventional products
7	Fruit and vegetable free from pesticides are more natural
8	Production process of pesticide-free fruits and vegetables is environmentally friendly

Table 3 Statements of consumer's safer shopping criteria in buying fruit and vegetables

1	The size of the vegetable leaves is not large
2	Healthy (free of harmful chemicals)
3	Not having the same sizes (for fruits)
4	Minor corrosion in the fruit
5	Product's freshness

Table 4 Statements of respondent's friendly attitude for environment index

1	I prefer to use recyclable products
2	Current methods of crop production cause environmental degradation
3	Destruction of the environment is irreversible unless action is taken
4	The environment is unique and should be supported with all facilities
5	The country is currently facing more serious problems than nature, such as unemployment and poverty
6	I am willing to give up some of my facilities to protect the environment
7	I am not willing to pay for environmental protection
8	Pesticide consumption should be stopped or minimized because of damage to the environment

At the second part of questionnaire, we tried to elicit respondents' demographic characteristics, like: age, family size, monthly income, educational level, gender, and matrimony status. Also, we asked them to determine if they had any kids younger than 10 years old, or a person with special disease in their family.

3 Results

The results of the survey on "Respondent's knowledge of PFFV Index" are summarized in Table 5. It shows that 60% of respondents were "agree" and "strongly agree" with these products' characteristics. Also 24% of them were "disagree" and "completely disagree" with the presented statements. The mean "Respondent's knowledge of PFFV Index" is about 3.4 and its minimum and maximum values are 1 and 5, respectively. From the results, it is clear that the majority of the consumers have acceptable and moderate knowledge about pesticide-free fruits and vegetables. However, only a small percentage (24%) is unaware of these products.

Table 5 Distribution of AWR variable's responses among households of Marand City

Responses	Frequency	Relative frequency	Cumulative frequency
Strongly disagree	41	10	10
Disagree	54	14	24
Indifferent	65	16	40
Agree	156	40	80
Strongly agree	78	20	100
Sum	394	100	–

Table 6 Distribution of shop variable's responses among households of Marand City

Responses	Frequency	Relative frequency	Cumulative frequency
Strongly disagree	21	5	5
Disagree	51	13	18
Indifferent	92	23	41
Agree	165	42	83
Strongly agree	65	17	100
Sum	394	100	–

The results of the respondents' survey of the "safer shopping criteria in buying fruit and vegetables" are presented in Table 6, which shows that 59% of respondents were "agree" and "strongly agree" with healthy shopping criteria and 23% of them reacted indifferently to them. The mean of the index among the sample population was 3.5 and the minimum and maximum values were 1 and 5, respectively. The results indicate the moderate importance of these criteria among the respondents.

Table 7 shows the summary results of "Respondent's Friendly attitude for environment Index." As it reveals, most of the respondents were "agree" and "strongly agree" with the friendly statements that were brought to them, and only 9% of them thought otherwise. The mean of this index is 3.9 and the minimum and maximum values are 1 and 5, respectively. This highlights that the respondents have high priority for environmental protection.

Table 8 briefly shows the descriptive statistics of respondents' demographic characteristics. Moreover, based on the results, the sample members are middle-income and middle-aged. Also their family size is average. Furthermore, they are mostly males (58%) and married (83%). Seventy percent of them have children younger than 10 years old (45%) or people with special disease in the household (25%). Descriptive results shows that consumers' mode of education was 4 (i.e., Diploma).

Table 7 Distribution of ENV variable's responses among households of Marand City

Responses	Frequency	Relative frequency	Cumulative frequency
Strongly disagree	6	1	1
Disagree	30	8	9
Indifferent	75	19	28
Agree	184	47	75
Strongly agree	99	25	100
Sum	394	100	–

Table 8 Descriptive statistics of respondents' demographic characteristics

Variable name	Mean*	SD	Min	Max
INC (Rials)	6020000	2355000	700000	30000000
Age	39	10.4	20	70
FDIM	3.8	3.7	1	10
EDU	4.8	1.39	1	8
Gender	1	0.6	0	1
Matri	1	0.8	0	1
SPCL	1	0.4	0	1

* For the binary variables, the mode of the variable is shown instead of its mean

Estimated results of ordered-probit model are presented in Table 9. The chi-square (χ^2) statistic, which is significant at the 1% level, points out estimated model's satisfactory explanatory power. Also, the scaled (60%) shows the model's overall ability for providing accurate prediction for the dependent variable. The results of Brant test show that the parallel regressions assumption is not violated in the estimated model.

According to Table 9, age is not an important factor that affects the consumer's awareness of PFFV. This means sample consumers' knowledge of PFFV, does not follow any age pattern, the same result was reached in similar studies (Hayati et al. 2017; Haghjou et al. 2013; Cranfield and Magnusson 2003; Boccaletti and Nardella 2000). Also, a significant and positive sign of income shows its important effect on respondent's knowledge of PFFV. This could be due to the quality search on richer consumer, meaning because of better well off; they search and ask about organic and safer products even though they might be more expensive. Similar results could be found in the literature (i.e., Hayati et al. 2017; Haghjou et al. 2013; Oni et al. 2005; Loureiro and Umberger 2004; Roitner-Schobesberger et al. 2008). According

Table 9 Estimation results of ordered probit model

Variable	Estimated coefficient	Standard error	Z-ratio
Constant	-3.93^{***}	0.39	−10.04
INC	0.123^{***}	0.31	3.90
Age	0.960^{ns}	0.80	1.19
SShop	0.846^{***}	0.91	9.24
ENV	0.572^{***}	0.82	9.91
EDU	0.138^{***}	0.60	2.29
Gender	-0.807^{***}	0.13	−6.19
Matri	0.568^{***}	0.20	2.74
SPCL	0.778^{***}	0.20	3.84
Log likelihood: −502.12, $R_s^2 = 0.58$, χ^2: 394.51^{***}			

***: $P < 0.01$, **: $P < 0.05$, *: $P < 0.1$, ^{ns}: non-significant to the estimated model, respondents with higher concerns about healthy fruits and vegetables, besides those with higher friendly attitudes toward environment, have more knowledge about PFFV. This results ties well with previous studies (Hayati et al. 2017, Haghjou et al. 2013, Roitner-Schobesberger et al. 2008, Williams and Hammitt 2000, Underhill and Figueroa 1996). According to the model, respondents with higher educational levels are those who are more aware of pesticide-free fruits and vegetables, this would mention importance of education in promoting community's concern about safer food products. Finally, results reveal that females and married respondents along with consumers who have kids or patient members in their families have more knowledge of pesticide-free fruits and vegetable, This indicates the higher importance they place to food safety and their family health. Some resembling studies also reached to this result (i.e., Hayati et al. 2017; Haghjou et al. 2013).

In the ordered-probit model, like all of the nonlinear models, marginal effects are used to investigate the effect of changing a specific independent variable on the dependent one. The sum of variable's marginal effects across all of the five categories should be zero. To interpret the marginal effects for non-binary variables, we should consider the other factors fixed; equal to the degree of marginal effect, one unit change in the independent variable would result in an increase or decrease in the predicted probability of the dependent variable, while in the case of a binary variable, the marginal effect means changing in the predicted probability according to the condition of the respondent (whether he/she falls into the particular category or not). Table 10 shows the results from marginal effects calculation of the probit model.

The marginal effects sign and values for the INC variable indicates that by the increase of respondents' income, the probability of being aware and fully aware of PFFV increases, while the probability of the first two lower classes of AWR (low and very low), declines. Therefore, it can be said that persons with higher incomes, are consumers who have higher awareness of pesticide-free fruits and vegetables. The highest positive effect was observed at level 5 (the highest awareness = strongly agree) and the highest negative effect at level 2 (low awareness = disagree). For

Table 10 Marginal effects variables from the ordered probit model

Variable	AWR = 1	AWR = 2	AWR = 3	AWR = 4	AWR = 5
Constant	0	0	0	0	0.0
INC	−0.0001	−0.0004	0.0002	0.0002	0.0004
Age	−0.0009	−0.0028	0.014	0.019	0.004
SShop	−0.078	−0.249	0.125	0.168	0.038
ENV	−0.0525	−0.168	−0.084	−0.114	0.026
EDU	−0.012	−0.041	0.020	−0.027	0.006
Gender	0.061	0.232	−0.073	−0.171	−0.052
Matri	−0.071	−0.145	0.112	0.094	0.017
SPCL	−0.043	−0.222	0.023	0.176	0.069

example, for the level 4 of AWR, marginal effect of INC variable is about 0.0002, which means with one unit of income rising, and other conditions remaining constant, among those who are at fourth level of AWR, their awareness increases about 0.0001. The same explanation could be made for the EDU, Shop, and ENV variables; which means as the educational level, also respondents' care about a healthy shop as well their friendly attitudes' toward environment rise, the probability of placing in the three higher classes of awareness about PFFV increases, while the probability of being in the two lower classes of awareness decreases. The results from variable gender show that being male would decrease the probability of being in the higher levels of awareness about PFFV; while being female is vice versa. Finally, being married, as well as having kids younger than 10 years old or having patient member in the family, results in being in the higher levels of awareness about pesticide-free fruits and vegetables.

4 Conclusion and Recommendations

The positive and significant effect of respondents' income with their awareness of PFFV was expected, because people, who are more financially capable, in addition to the quantity of products they purchase, consider their quality and study about health. In this regard, the empowerment of low-income people to buy healthier food products through special subsidies which is a form of income distribution, can lead deprived and poorer people toward consumption and recognition of these types of products and the overall correction of consumption pattern of the community.

Considering the significant relationship of environmentally friendly tendencies with awareness about pesticide-free fruits and vegetables, also regarding the fact of information and advertising weaknesses about safer food products in the community, supporting for formation of nongovernmental organizations (NGO)s which are activating about environmental conservation and sustainable agriculture along with encouraging them to increase community's information about such issues, would be an influential move.

The positive and significant effect of the healthy Shopping Index indicates that the higher the level of awareness, attention, and importance of the health criteria during shopping, results in the greater the consumer's interest in pesticide-free products. This would be a clue for the policy makers to do some plans for attracting consumers during the shopping and changing their purchase patterns; e.g., by installing or distributing leaflets in shopping malls and explaining about pesticide-free and healthy fruits and vegetables and their features.

According to the results, educational level of respondents showed a significant and positive effect on their awareness of PFFV. This suggests that educational systems must first be assisted to institutionalize and familiarize society with the culture of consumption and production of sustainable agricultural products. Considering this fact, targeted trainings at all levels of educational system for the cauterizing of consumption and production of these products in society can be considered as an important step toward cultural development and familiarization of the society with these types of products.

The results showed that females were more aware of the pesticide-free fruits and vegetables, which could be due to this fact that they have much more free time to study about health and safety of products they purchase. Therefore, increasing awareness of males in the workplace through training classes can also inform this group of consumers about healthier food products and also PFFV. Moreover, since female respondents are more aware of PFFV, and most likely they are the main buyers of fruits and vegetables, doing some attractive things, such as providing ladies' favorite sample products with information brochures about PFVV could have a positive result to get their attentions.

Given the positive and significant effect of increased awareness with the presence of people with specific illnesses or children under the age of ten in the households, it indicates that people facing such problems may seek to increase their awareness and improve their dietary patterns and use of healthy foods. Due to what was mentioned, targeted production of safer and pesticide-specific food products which is specialized for these groups of consumers, specific and targeted packaging of such products, putting information on packaging that they are suitable for the use of people with specific diseases or children, are through recommendations of the present study.

References

Alizade A, Javanmardi J, Abdollahzade N, Liaghati Z (2008) Consumers' awareness, demands and preperences for organic vegetables: a survey study in Shiraz, Iran. In: 16th IFOAM organic world congress, Modena, Italy, June 16–20

Ahmad SNB, Juhdi N (2008) Consumer's perceptions and purchase intentions toward organic food products: an explorative study on attitudes of Malaysian consumers. In: 16th annual conference on pacific basin finance, economics, accounting and management, Brisbane Australia, July, 2–4

Boccaletti S, Nardella M (2000) Consumer willingness to pay for pesticide-free fresh fruit and vegetables in Italy. Inter Food Agribus Mng Rev 3(3):297–310

Boxall P, Cash S, Wismer W, Muralidharan V, Annet L (2007) The role of sensory experiences and information on willingness to pay for organic wheat bread. J For Econ 27:16–29

Cowan C, Carthy M, Riodan N (2000) Irish consumers' willingness to pay for safe beef. J Consum Res 32:146–153

Chinnici G, D'Amico M, Pecorino B (2002) A multivariate statistical analysis on the consumers of organic products. Brit Food J 104(3/4/5):187–199

Cranfield J, Magnusson E (2003) Canadian consumer's willingness-to-pay for pesticide free food products: an ordered probit analysis. Inter Food Agri Mng Rev 6(4):13–30

Darby K, Batte M, Ernst S, Roe B (2006) Willingness to pay for locally produced foods: a customer intercept study of direct market and grocery store shoppers. Selected Paper Prepared for Presentation at the American Agricultural Economics Association Annual Meeting, July 23–26, Long Beach, California

Ghorbani M, Mahmoudi H, Liaghati H (2007) Consumers' demands and preferences for organic food: a survey study in Mashhad, Iran. Poster presented at the 3rd QLIF congress: improving sustainability in organic and low input food production system, University of Hohenheim, Germany, March 20–23

Gold M (2018) Sustainable agriculture: information access tools. United States Department of Agriculture, Alternative Farming Systems Information Center
Greene W (2005) Econometric analysis. Macmillan, NewYork
Erfanmanesh M, Afyouni M (2007) Environmental pollution, 4th edn. Ardakan Publication, Esfahan, Iran (In Persian)
Haas R, Sterns J, Meixner O, Nyob DI, Traar V (2013) US Consumers' perception of local and organic food: an analysis based on means-end chain analysis and word Association. Institute of Marketing & Innovation, Department of Economics and Social Sciences. http://centmapress.ilb.unibonn.de/ojs/index.php/proceedings/article/download/327/309
Haghjou M, Hayati B, Pishbahar E, Mohammadrezaei R, Dashti G (2013) Factors affecting consumers' potential willingness to pay for organic food products in Iran: case study of Tabriz. J Agri Scie a Tech 15(2013):191–202
Harper GC, Makatouni A (2002) Consumer perception of organic food productions and farm animal welfare. Bri Food J 4:287–299
Hayati B, Haghjou M, Pishbahar E (2017) Effecting factors on consumers' willingness to pay a premium for pesticide-free fruit and vegetables in Iran. MOJ Food Proc Tech 4(5)
Howard A, Berry W (2008) The soil and Heath: a study of organic agriculture. The University Press of Kentucky
Iranian Statistic Center (2012) Annual report of statistic (In Persian)
Loureiro M, Umberger W (2004) A choice experiment model for beef attributes: what consumer preferences tell us. Selected Paper Presented at the American Agricultural Economics Association Annual Meetings, Denver, Colorado, August 1–4
Mahmoudi H, Mahdavi-Damghani A (2010) Organic agriculture in Iran-country report. http://www.organicworld.net/iran0.html. IFOAM. International Federation of Organic Agriculture Movements. http://www.ifoam.org/
Millock K, Hunsen L, Wier M, Andersen L (2005) Willingness to pay for organic foods in Denmark. J Political Econ 75:132–157
Oni O, Oladele O, Inedia O (2005) Consumer willingness to pay for safety labels in Nigeria: a case study of potassium bromate in bread. J Cen Euro Agri 6(3):381–388
Pishbahar E (2018) Econometrics (with the application of most frequently used software packages for econometrics). Nore-elm Publications, Iran (In Persian)
Rodriguez E, Lacaze V, Lupin B (2007) Willingness to pay for organic food in Argentina: evidence from a consumer survey. Papers prepared for 105th EAAE Seminar, Bologna, Italy
Roitner-Schobesberger B, Darnhofer I, Somsook SR, Vogl C (2008) Consumer perceptions of organic foods in Bangkok. Thai Food Pol 33(2):112–121
Sachs C, Blair D, Ritcher C (1987) Consumer pesticide concerns: a 1965 and 1984 comparisons. J Consum Aff 2:96–107
Shafie FA, Rennie D (2012) Consumer perceptions towards organic food. Procedia Soc Behav Sci 42(2012):360–367
Tison AM (2012) A study of organic food consumers' knowledge, attitudes and behavior regarding labor in organic farms. Consumer Knowledge of Labor in Organic Farms. http://nature.berkeley.edu/classes/es196/projects/2012final/TisonA_2012.pdf
Torjusen H, Nyberg A, Wandel M (2010) Organic food: Consumer's perceptions and dietary choices a survey from the Stange and Hamar region. English Summary, Statens Institute for Forbruksforskning
Underhill S, Figueroa E (1996) Consumer preferences for non-conventionally grown produce. J Food Distri Res 27(2):56–66
Williams PR, Hammitt JK (2000) A comparison of organic and conventional fresh produce buyers in the Boston area. Risk Anal 20(5):735–746
Yiridoe EK, Bonti-Ankomah S, Martin RC (2005) Comparison of consumers' perceptions and preferences toward organic versus conventionally produced foods: a review and update of the literature. Renew Agr Food Syst 20:193–205

Maryam Haghjou is a lecturer in Department of Agricultural Economics at University of Tabriz. She holds her B.S., M.Sc. and Ph.D. from Department of Agricultural Economics, University of Tabriz. She has taught a number of courses on Natural Resource Economics and Microeconomics. Her research and publication has focused on Sustainable Development, Environmental Management and Natural Resource Management. She has publications in Journal of Agricultural Science and Technology (JAST).

Babollah Hayati is a Professor in Department of Agricultural Economics at University of Tabriz. He was dean of Faculty of Agriculture during 2015–2018. He holds a B.Sc. in Agricultural Economics at University of Tehran and a M.Sc. and Ph.D. in Natural Resource Economics at Tarbiat Modares University. His areas of special interest are Natural Resource Economics, Sustainable Development Economics and Microeconomics. His recent publications have appeared in numerous journals including the Journal of Agricultural Science and Technology (JAST) and Engineering Sustainability.

Esmaeil Pishbahar is Associate Professor of Agricultural Economics at University of Tabriz, Iran. He holds a B.Sc. in Agricultural Economics from University of Tabriz and a M.Sc. in Agricultural Economics from University of Tehran. He did his Ph.D. in Science Economics at departments of Economics and Management, University of Rennes 1, France. His areas of interest and research are Applied Econometrics, Agricultural Risk Management and Insurance, and International Trade. His teaching area are Advanced Econometrics, Mathematical Economics, and Macroeconomics at under- and postgraduate levels. He has over 100 publications in journals and chapters in books.

Energy Use in Agriculture

The Relationship Between Economic Growth, Energy Consumption, and CO_2 Emissions

Parisa Pakrooh and Esmaeil Pishbahar

Abstract Iran's economic growth over the last decades has been accompanied by many structural shocks and volatilities, which affect other related variables such as energy consumption and CO_2 emissions. Therefore, this study analyzes the relationship between economic growth, energy consumption (including oil, gas, coal, renewable, and electricity), and CO_2 emissions. The important differences of our study are in examining the dynamic relationship among variables concerning volatilities and structural shocks, because of that the TVP-VAR method applied to this aim during 1978–2015. This method helps us to understand the kind of relationship, response of variables to each other's volatilities and structural shocks, and also the size of response, which is necessary in making accurate policies. Also, stationary, co-integration and causality test used, and then the response and the size of responses to structural shocks and fluctuations of variables are analyzed with TVP-VAR model. The results indicate that there is a bidirectional relationship between energy consumption and GDP, and directional relationships between energy consumption and CO_2, also GDP and CO_2. The results TVP-VAR model shows that the dynamic Kuznets' theory exists, and the size of the response of energy consumption and CO_2 emissions to GDP changes has been lager than other statues. So policymakers must reduce the energy consumption and CO_2 emissions reactions by replacing renewable energies, which can help to reduce the dependency of sectors to fossil energies and GDP as well as CO_2 reduction.

Keywords Energy · Environment · Iran · Stochastic volatility · Time-varying parameter vector autoregressive model

P. Pakrooh (✉) · E. Pishbahar
Department of Agricultural Economics, University of Tabriz, Tabriz, Iran
e-mail: parisa.pakrooh@tabrizu.ac.ir

E. Pishbahar
e-mail: pishbahar@tabrizu.ac.ir

M. Rashidghalam (ed.), *The Economics of Agriculture and Natural Resources*, Perspectives on Development in the Middle East and North Africa (MENA) Region,
https://doi.org/10.1007/978-981-15-5250-2_10

1 Introduction

Over the past 200 years, economic growth worldwide has a sharp increase trend. It is necessary to achieve such high economic growth, use many inputs such as energy, so along with economic production, we have pollution too. A comparison trend of pollution emission and economic growth shows that there was a slight upward trend until 2000. Still, the situation has continued to change, and the emission of pollutants has increased until now; also it has caused many environmental problems (Wolde et al. 2016).

Global warming and climate change have always been one of the crucial environmental problems in the last two decades due to the increased size of greenhouse gas emissions; meanwhile, CO_2 emission is more important because of its high volume of greenhouse gases. Recent studies showed that fossil energy consumption is one of the most important contributors to the CO_2 emissions, which is increasing as a result of industrial activities (Ozturk and Acaravci 2010; Alam et al. 2011; Pakrooh and Pishbahar 2019; Zhang and Cheng 2009).

According to the Energy Information Administration (EIA), Iran is fourth rank in oil reserves and second rank in natural gas reserves in the world, so it called energy superpower country. Due to that, 98% of energy demand in Iran's economic sectors (including industry, transportation, agriculture, household, and commercial) are provided from fossil fuels, and renewable energy has almost less share in energy mix. In 2014, Iran's Ministry of Energy stated that about 2523.5 million barrels of energy was used in economic sectors, which includes 1259 million barrels oil, 8.537 million barrels renewable combustible sources, 9.3 million barrels coal, 1231 million barrels natural gas, 8.8 million barrels solar and wind power, and 2.5 million barrels nuclear power (Ministry of Energy 2018). However, fossil fuels generate toxic and polluting air, acid rain, and consequently contamination of rivers, lakes, and groundwater and eventually rising CO_2 levels in the atmosphere of Iran. Simultaneously in 2014, 602 million tons of CO_2 released due to the high consumption of fossil fuels and incomplete combustion (Sharifi et al. 2012). Hence, Iran became the ninth most polluted country in the world, and some of the megacities, such as Ahvaz, was the first polluted city in the world. It seems that due to the high size of fossil energy consumption (oil, gas) in the economic sectors to production and achieve sustainable economic, makes Iran as one of the most polluted countries in the world that can contribute to global warming and climate change (EIA 2015; Environmental Assessment Agency 2016).

Environmental issues such as CO_2 emissions caused by energy-related economic activities are one of the main goals of many policymakers and governments. Rapid economic growth, due to the increasing use of energy, usually causes severe environmental damages. Hence, there is often a conflict between economic growth and environmental pollution. International organizations try to reduce environmental pollution and climate change through agreed protocols such as Kyoto in 1997 (Ozturk and Acaravci 2010). One of the recent major international agreements on this matter is the Copenhagen Protocol in 2009, which emphasizes the reducing greenhouse gas emissions such as CO_2 to protect the environment (Alam et al. 2011). According to the Brent land report, "Sustainable Development" is an extension that addresses the current needs of the world, which designed to sustain economic, social, and environmental development, and the most important elements of sustainable development are energy resources. Therefore, having the right energy mix is the main factor for sustainable economic growth, social welfare, quality of life, and community security (Barimani and Kaabinezhad 2014; Hosseininasab and Paykari 2012).

The consequences of excessive consumption of energy and environmental degradation have led researchers in recent decades to investigate the relationship between pollutant emissions and economic growth. On the other hand, investigating the relationship between energy consumption and economic growth due to global warming and climate change is one of the crucial problems of the last two decades worldwide. Hence, analyzing the relationship between economic growth, energy consumption, and CO_2 emissions is useful to find out how are the relationships, and the empirical results help policymakers and planners to move forward sustainable development.

The rest of the paper organized as follows. We review the literature in Sect. 2, and afterward, in Sect. 3, we present the methodology. Section 4 applies the methodology to the Iranian case study. Finally, Sect. 5 gives conclusions.

2 Literature Review

On the one hand, there is a possible relationship between pollution and energy consumption, and the relationship between energy consumption and economic growth has made the issue of the effects of pollution on economic growth to be important. Given the relationship between economic growth and environmental degradation, in recent years, there has been a debate between supporters of economic growth, which means that economic growth requires higher energy consumption, which may lead to environmental degradation. Accordingly, environmentalists believe that economic growth should slow down to improve the quality of the environment and preserve it. On the other hand, economic growth supporters believe that, according to Kuznets, economic growth can accompany a reduction in pollution. Today, the Kuznets' curve uses as a means to explain the relationship between environmental quality and per capita income. This U-inverse relationship is called the Kuznets' environmental curve between economic growth and pollution measurements (environmental quality). In the early stages of economic growth, awareness of environmental problems is low, and environmentally friendly technologies are not available. Hence, with increasing revenue to reach the threshold level, degradation of the environment increases, but after that, the quality of the environment will improve as well as improving per capita income. This connection is known as EKC (Environmental Kuznets' Curve) (Kuznets 1955). Based on the theory, increasing economic growth is providing resource intensification, and put the agricultural sector under high production pressure. In higher developmental stages, restructuring of the industry, information services along with increasing environmental awareness, the requirement for

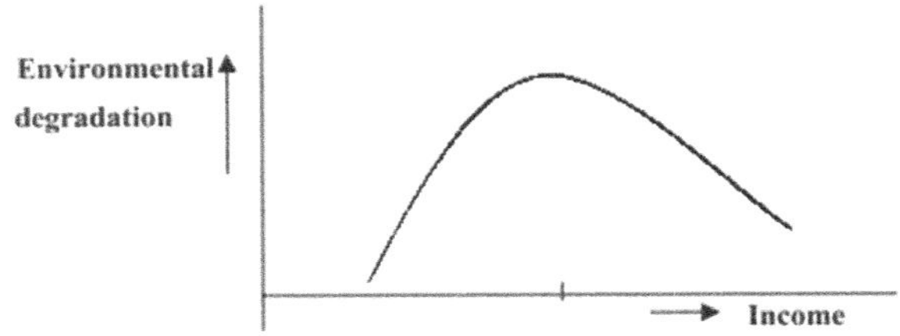

Fig. 1 Kuznets' curve. *Source* Dinda (2004)

environmental regulations, better technology, and higher environmental costs will reduce environmental degradation. Finally, when revenue reaches the ECC point, move toward improving environmental quality (Fig. 1) (Dinda 2004).

The relationship between economic growth and environmental pollution is one of the issues highlighted over the past decades. Recently, Kuznets' environmental curve was investigated by Fetras and Nasrindoost (2009), Hosseini Nasab and Paykani (2011), Fodha and Zaghdoud (2010) and Oganesyan (2017), using per capita income and environmental quality indices. It is clear that the relationship between economic growth, energy consumption, and environmental degradation as one of the important issues have considered in several studies discussed in the Table 1.

By examining studies in Table 1, it seems that over the past decades, concerns upon the economic growth, fossil energy consumption, and pollutant emissions have increased in the world. Since societies face environmental pollution as one of the most important issues that need to address a variety of studies. As can be seen, studies in the field of research divided into three categories. Primary, surveys the impact of **environmental quality and energy consumption on economic growth**, which is analyzed by Kuznets' hypothesis in time series and panel data using VAR, VECM, and ARDL models. The second group examined the causality between economic growth, energy consumption, and emission of pollutants to **determining the direction of causality**. The third group studies analyze the impact of **economic growth on energy consumption and environmental quality** using VAR, VECM, and ARDL models. There are few studies on the dynamic relationship and stochastic volatility effects between economic growth, energy consumption, and environmental quality. Only Mezghani and Haddad (2016), employ the TVP-VAR approach with stochastic volatility to examine the dynamic relationship between Saudi Arabia's economic growth, electricity consumption, and the quality of the environment noted. In sum, most studies are weak in terms of considering the dynamic relationship between variables and the effects of dynamic stochastic volatility with stochastic fluctuations of the variables. Therefore, in this study, after determining the causality direction, we investigate the relationship among economic growth, energy consumption, and CO_2 emission by using the TVP-VAR approach, which considers the stochastic volatility of variables from 1979 to 2015. We applied the TVP-VAR model; according to Nakajima (2011) state the TVP-VAR model is a functional model for estimating the dynamic relationship between the variables, after all, it determines changes between variables over time.

Table 1 Summary of related studies

Author(s) and years	Indicator	Method and study period	Conclusion
Lee (2006)	CO_2	VAR 1965–2001	The causal relationship between energy consumption and GDP for the industrialized countries such as England, Germany, and Sweden has shown that there was an unreliable relationship but in the United States, Canada, Belgium, the Netherlands, and Switzerland was a direct relation from energy consumption to GDP and finally, in Japan, Italy, and France are indirect
Zhang and Chen (2009)	CO_2	VAR 1960–2007	There was an indirect causal relationship between China's economic growth and energy consumption and energy consumption with pollutant production in the long run
Fetras and Nasrindoost (2009)	CO_2	VAR 1980–2004	The purpose of this study were to investigate the EKC hypothesis in four scenarios: (1) equality of economic growth rate with per capita income, (2) equality of growth criterion with energy consumption, (3) pollution criterion is water pollution, (4) pollution criterion is air pollution, the results show that the Kuznets hypothesis accepted in Iran

(continued)

Table 1 (continued)

Author(s) and years	Indicator	Method and study period	Conclusion
Fodha and Zaghdoud (2010)	CO_2 and SO_2	VAR 1961–2004	There was a long-term relationship between per capita income and gas emissions in Tanzania, and the Kuznets' environmental hypothesis is accepted. But the relationship between income and pollutants was indirect in the short and long terms
Moghaddasi and Rahimi (2010)	PSI	Panel, FE 2004–2009	The Kuznets' hypothesis was accepted in the provinces of Tehran, Khuzestan, and Khorasan, and with increasing production of these provinces, emissions of pollutants and SPM have increased
Ozturk and Acaravci (2010)	CO_2	ARDL 1968–2005	Kuznets' hypothesis was not accepted in Turkey, but there was a long-term relationship between the variables. The causality test showed the causality relation between energy consumption and economic growth
Alam et al. (2011)	CO_2	VAR 1971–2006	There was a causal relationship between energy consumption and pollutant emissions in the long run, but there is no evidence that energy consumption and emissions are related to economic growth in India

(continued)

Table 1 (continued)

Author(s) and years	Indicator	Method and study period	Conclusion
Hosseininasab and Paykari (2012)	CO_2 and BOD	Panel, RE 1980–2000	The Kuznets' hypothesis was accepted for both air and water pollution during the mentioned period for developed countries, but in developed countries, only the water pollution index was consistent with this hypothesis
Mahdavi and Amirbabaei (2013)	CO_2	ARDL 1970–2007	The coefficient of financial development index with a negative sign was in the model, which indicates the relationship between the financial development index and the publication. Therefore, it would be possible to expect an increase in emissions at high levels of financial development
Nazari et al. (2014)	CO_2	GMM 1973–2013	Economic growth, energy consumption, and the number of vehicles have a positive and significant effect on environmental pollution, but the degree of openness of the economy, temperature, and rainfall are negatively related to environmental pollution

(continued)

Table 1 (continued)

Author(s) and years	Indicator	Method and study period	Conclusion
Lim et al. (2014)	CO_2	VECM 1965–2012	There was a bidirectional causal relationship between oil consumption and the growth of the Philippines and from oil consumption to pollution emissions, but it has not been seen as a relation from pollutant emissions to economic growth
Mahdavi Adeli and Nazari (2015)	CO_2	GMM 1973–2013	The effect of energy consumption and environmental pollution on economic growth was positive and significant and also the effect of economic growth on energy consumption was positive and significant. Finally, energy consumption and economic growth have a significant effect on pollution
Chindho et al. (2015)	CO_2	ARDL 1960–2013	Pollutant emissions have a significant and positive effect on GDP in Nigeria, and with increasing levels of pollution, GDP is increase. On the other hand, energy consumption in the short run has a negative impact on GDP

(continued)

Table 1 (continued)

Author(s) and years	Indicator	Method and study period	Conclusion
Mezghani and Haddad (2016)	CO_2	TVP-VAR 1971–2010	Fluctuations in electricity consumption in Saudi Arabia have had a negative impact on oil-GDP and CO_2 but have had a positive impact on non-oil sector GDP. Also, oil and non-oil GDP fluctuations have a positive effect on electricity consumption and emissions
Alizadeh and Golkhandan (2016)	Energy consumption	Panel 1990–2011	The results of this study indicate the energy consumption has a positive effect on the economic growth in OPEC countries in the long and short terms. There is also a bidirectional relationship between economic growth and energy consumption in the short and long term
Tekaleg et al. (2016)	CO_2	VAR 1970–2011	Ethiopia's oil shock hurts growth and there is no relationship between energy consumption and economic growth, but economic growth is positively correlated with CO_2 emissions

(continued)

Table 1 (continued)

Author(s) and years	Indicator	Method and study period	Conclusion
Salmanpour et al. (2017)	CO_2	ARDL 1986–2016	The results show a relationship between per capita income and environmental pollution. Therefore, Kuznets' environmental hypothesis is accepted in Iran and there was a positive relationship between oil consumption, and environmental contamination
Oganesyan (2017)	CO_2	Linear model 1980–2012	The elasticity of CO_2 relative energy consumption in BRIC member was 0.6, and the economic growth rate elasticity relative energy consumption was 1.74. The causality test shows that there is a relationship between energy consumption and pollution emissions
Airkhizi et al. (2018)	CO_2	Panel 1995–2015	The results showed that the Kuznets' hypothesis was accepted in the studied countries. The coefficient of GDP growth and its squared was respectively had positive and a negative effect. Also, energy consumption has a positive and statistically significant effect on gas emissions

*SPM or suspended particulate matter such as dust, fumes, mist, and smoke. The concentration of these in and near the urban areas causes severe pollution to the surroundings

*GMM or generalized method of moments is a generic method for estimating parameters in statistical models

3 Materials and Methods

VAR (vector autoregression model) is a model that uses to show dependencies between more than one-time series variables, and predict the future amount of series. This model is one of the most important practical tools in the econometric modeling, in which a TVP-VAR model with time-varying parameters and stochastic volatility proposed by Primiceri (2004). In this model, using time-varying variables is practical and assumed that all parameters are a stochastic walk process. So, it allows existing temporal and permanent changes in the parameters, and IRF functions show the dependent variables changes by other variables. TVP model is as Eq. (1) (Nakajima 2011; Primiceri 2004):

$$y_t = x_t'\beta + z_t'\alpha_t + \varepsilon_t, \varepsilon_t \sim N(0, \sigma_t^2)\ (t = 1, \ldots, n) \tag{1}$$

Time-varying parameters of Eq. (1) are in the form of Eq. (2) (Nakajima 2011; Primiceri 2004):

$$\alpha_{t+1} = \alpha_t + u_t, u_t \sim N(0, \textstyle\sum), (t = 0, 1, \ldots, n) \tag{2}$$

And stochastic volatility is written as follows.

$$\begin{aligned} \sigma_t^2 &= \gamma \exp(h_t) \\ h_{t+1} &= \varphi h_t + \eta_t, \eta_t \sim N(0, \sigma_\eta^2), (t = 0, 1, \ldots, n) \end{aligned} \tag{3}$$

where y_t is a response vector, x_t and z_t are the convergence vectors, β is a vector of static coefficient, α is a vector of dynamic coefficient, and h_t is stochastic volatility, also assumed $\alpha_0 = 0, u_0 \sim N(0, \sum_0), \gamma \geq 0$, and $h_0 = 0$ (Nakajima 2011; Primiceri 2004).

The Eq. (1) has two convergence components: the first component, β, is the static coefficients, and the second component, α, is related to the coefficients of time variables. The effect of x_t on y_t assumed to be independent of time. But the effect of z_t on the y_t is variable over time. Also, the ε_t has a normal distribution with time-varying σ_t^2 variance. In Eq. (2), α is formulated to have temporary and permanent changes such as structural shock and minor variations (Nakajima 2011; Primiceri 2004).

Stochastic volatility plays a crucial role in the TVP model, most of researchers use this model in different field of economy but Primiceri (2004) used this model in their macroeconomic study. Stochastic volatility makes the estimation difficult due to the maximum exponential function, so the Monte Carlo Markov Chain (MCMC) method used to estimate the model. In estimating the TVP models, there are several reasons for using the MCMC model. First, the maximum likelihood function is solvable because the model has nonlinear equations of stochastic volatility. Second, the application of this model is simultaneously in parameters and static variables, provided inference and conclusion for state variables with uncertainty in parameters (Nakajima 2011; Primiceri 2004).

The MCMC method is based on the Bayesian inference, and aims to determine the joint pre-distribution of the parameters with density probability pre-distribution that researchers intend to perform. In the Bayesian inference, the pre-density defined as a vector $\pi(\theta)$ for the parameters, and the maximum likelihood function specified $f(y|\theta)$ for the data $y = \{y_1, y_2, \ldots, y_n\}$. The inference based on the pre-distribution, introduced by $\pi(\theta)$ Bayesian theory, is in the form of Eq. (4) (Nakajima 2011; Primiceri 2004; Barnett et al. 2014).

$$\pi(\theta|y) = \frac{f(y|\theta)\pi(\theta)}{\int f(y|\theta)\pi(\theta)d\theta} \tag{4}$$

The TVP-VAR model has several advantages over different VAR models for two major reasons: (1) It has high flexibility in entering variable parameters over time; (2) This model allows time variations to be independent in any equation of the VAR model. Therefore, the TVP-VAR model with the basic structure of the VAR is defined as follow:

$$Ay_t = F_1 y_{t-1} + \cdots + F_s y_{t-s} + u_t, (t = s+1, \ldots n) \tag{5}$$

where, y_t is a vector of observable variables, and A, $F_1 \ldots F_s$ is a matrix of coefficients. Distribution of u_t is such a structural shock which assumes $u_t \sim N(0, \sum\sum)$ so:

$$\sum = \begin{pmatrix} \sigma_1 & 0 & \ldots & 0 \\ 0 & \ldots & \ldots & \ldots \\ \ldots & \ldots & \ldots & 0 \\ 0 & \ldots & 0 & \sigma_k \end{pmatrix} \tag{6}$$

$$A = \begin{pmatrix} 1 & 0 & \ldots & 0 \\ \alpha_{21} & \ldots & \ldots & \ldots \\ \ldots & \ldots & \ldots & 0 \\ \alpha_{k1} & \ldots & \alpha_k, \alpha_{k-1} & 1 \end{pmatrix} \tag{7}$$

Now, we can rewrite the Eq. (5) in the reduced form of the VAR model:

$$y_t = B_1 y_{t-1} + \cdots + B_s y_{t-s} + A^{-1} \sum \varepsilon_t, \varepsilon_t \sim N(0, I_k) \tag{8}$$

where $B_i = A^{-1}F_i$ $(i = 1, \ldots, s)$, and $x_t = I_s \otimes (y'_{t-1}, \ldots, y'_{t-s})$. Finally, the model can write in the form of Eq. (9). All parameters of Eq. (9) are time-varying.

$$y_t = x_t \beta + A^{-1} \sum \varepsilon_t \tag{9}$$

Now with the addition of stochastic volatility, the TVP-VAR model defined as Eq. (10) (Nakajima 2011; Primiceri 2004; Barnett et al. 2014).

$$y_t = x_t\beta + A^{-1}\sum \varepsilon_t + u_{ht} \tag{10}$$

There are several ways to model the time-varying parameters in the Eq. (10), which are defined as Eq. (11) according to the study of Primiceri (2004):

$$\begin{aligned}
&\beta_{t+1} = \beta_t + u_{\beta t} \\
&\alpha_{t+1} = \alpha_t + u_{\alpha t} \qquad \begin{pmatrix} \varepsilon_t \\ u_{\beta t} \\ u_{\alpha t} \\ u_{ht} \end{pmatrix} \sim N\left(0, \begin{pmatrix} I & O & O & O \\ O & \sum\beta & O & O \\ O & O & \sum\alpha & O \\ O & O & O & \sum h \end{pmatrix}\right) \\
&h_{t+1} = h_t + u_{ht}
\end{aligned} \tag{11}$$

where, $t = s + 1, \ldots, n$, $\beta_{s+1} \sim N(\mu_{\beta_0}, \sum_{\beta_0})$, $\alpha_{s+1} \sim N(\mu_{\alpha_0}, \sum_{\alpha_0})$, and $h_{s+1} \sim N(\mu_{h_0}, \sum_{h_0})$ (Primiceri 2004).

The empirical models are

$$\begin{aligned}
Co_{2,t} &= \beta_1 Co_{2,t-1} + \alpha_1\sigma_{oil} + \alpha_2\sigma_{GDP} + u_{h1} + u_{h2} \\
Co_{2,t} &= \beta_1 Co_{2,t-1} + \alpha_1\sigma_{gas} + \alpha_2\sigma_{GDP} + u_{h1} + u_{h2} \\
Co_{2,t} &= \beta_1 Co_{2,t-1} + \alpha_1\sigma_{coal} + \alpha_2\sigma_{GDP} + u_{h1} + u_{h2} \\
Co_{2,t} &= \beta_1 Co_{2,t-1} + \alpha_1\sigma_{renewable} + \alpha_2\sigma_{GDP} + u_{h1} + u_{h2} \\
Co_{2,t} &= \beta_1 Co_{2,t-1} + \alpha_1\sigma_{electricity} + \alpha_2\sigma_{GDP} + u_{h1} + u_{h2}
\end{aligned} \tag{12}$$

where σ is the variance of independent variables (GDP as a proxy of economic growth and energy consumption). $B_{1t}, \ldots, B_{qt}$ is a (5 × 5) matrix of static parameters, β is a time-varying parameter of VAR model, and u_t is stochastic and structural volatility.

In this study, the TVP-VAR model is applied to analyze the relationship between variables GDP (billion Rials), consumption of various types of oil, natural gas, coal, renewable, and electricity energies (equivalent crude oil per million tons) and CO_2 (million ton) during 1978–2015. Data were collected from the Central Bank of the Islamic Republic of Iran, Energy Balance Sheet and Statistics Center of Iran.

Table 2 Unit root test results

Variables (LOG)	ADF I(0)	ADF I(1)	Zivot–Andrews I(0)	Zivot–Andrews I(1) and break point	Result of stationary	Optimal lag
GDP	−0.071	−3.606**	−3.35 1993	−5.45** 1989	I(1)	2
CO_2	−0.089	−5.23**	−4.015 1986	−5.28** 1992	I(1)	1
Oil	−1.92	−4.48**	−3.25 1987	−5.40 2009	I(1)	1
Gas	−1.54	−0.46**	−6.012** 1990	−4.90** 1989	I(1)	2
Coal	−1.035	−4.16**	−3.98 2005	−4.56 2009	I(1)	2
Renewable	−2.48	−6.25**	−6.74** 2005	−6.47 2005	I(1)	1
Electricity	−3.47	−16.87**	−2.99 1989	−19.86** 1985	I(1)	1

**The critical value of the Dickey–Fuller and Zivot–Andrews statistics is 2.96 and 4.80, respectively, at the level of 5%

4 Results

4.1 Unit Root, Co-integration, and Granger Causality Tests

The Dickey–Fuller and Zivot–Andrews tests are used to test the stationary. The results show that all variables were stationary at first difference, and breakpoints of all variables are determined. The unit root test results are reported in Table 2.

As discussed earlier, one of the objectives of this study is to investigate the effect of stochastic volatility of variables on other variables, which can be analyzed the variables' response functions. First, we identified the causality between variables, and the Johansson test is applied to determine the co-integration vectors between variables. The results of co-integration test are reported in Table 3. It suggests the existence of at most two co-integration vectors in the first equation, which includes CO_2, Oil, and GDP, but in other equations existence of at most one co-integration vector.

Since, in all equations, there is a maximum of one co-integration vector; it is possible to estimate the VAR model for all equations concerning the optimal lags. Also, to determine the relationships between variables the Granger causality test is applied. The results of the Granger causality test indicate that directional and bidirectional causality between variables which are as follows (Table 4).

A TVP-VAR model is applied to analyze the relationship between economic growth, energy consumption, and CO_2 emissions of Iran from 1979 to 2015, and the

Table 3 Co-integration test results

Equation	Number of cointegration vector	Results		
		Hypothesis	Resulted test	Critical value
CO_2, oil, GDP	2	r = 0 r = 1	30.74 **11.58	29.68 15.41
CO_2, gas, GDP	1	r = 0	**27.09	29.68
CO_2, coal, GDP	1	r = 0	**21.77	29.68
CO_2, renewable, GDP	1	r = 0	**20.33	29.68
CO_2, electricity, GDP	1	r = 0	**23.68	29.68

**Indicate rejection of null hypothesis at 5% level

empirical results of the model are as follow. Figure 2 show the autocorrelation function of the samples, the sample paths, and the pre-density of the selected variables in both states of static and dynamic. According to Fig. 2, the paths of the samples are constant, and the autocorrelations of the samples are decreasing; it means that the sampling method is efficient, and samples produced with a low level of autocorrelation. Also, the pre-density of the samples has a normal distribution. Based on Fig. 2, we found that there is no problem in sampling and distribution of samples, so the next steps of the model are in the following part.

Figure 3 shows the stochastic volatility of the GDP, which is decreasing during the period. From 1978 to 1985, the volatility of the GDP was high due to the revolution and the Iran–Iraq War. It happened because of the high volatility of oil prices, which is determined Iran's salary revenue and led economic growth to an unstable situation. From 1986 to 1990, the volatility is decreasing, but increased since 1991 due to the beginning of the construction period until 1996. We also saw the decreasing behavior from 1997 to now; also, these volatilities are still small. CO_2 emission has high volatility during 1978–2015. From the beginning of the period, CO_2 experienced fluctuation trend until 1981 but decreased until 1992 due to the Iran–Iraq War. Domestic productions in the interior industries related to fossil energies decreased, so CO_2 emission was also at the lower level. In the following years, due to the high population growth rate, high energy consumption, because the construction period raised the volatilities till 1996. But in the aftermath of these volatilities, the trend has been decreasing, which is visible in Fig. 4.

Figures 5, 6, 7 and 8 shows the stochastic volatilities in the different types of energy consumption in Iran during the studied period. Based on Fig. 5, among all kinds of energy consumption, the oil has a large size of volatility and a peak point in 1997 due to an increase in oil consumption for high economic growth purposes. The oil consumption increased from 1.31 million barrels in 1995 to 1.41 million barrels in 1997 and again decreased to 1.40 million barrels in 1999. For the rest of the years, the average consumption was less than the peak point. As shown in Fig. 6, natural gas consumption in recent years has a significant size of stochastic volatilities over the studied period when is related to first and second subsidizing targeting policies and sectors equips to use natural gas instead of oil. At the end of the period, natural

Table 4 Granger causality test results

Equation	Independent variables					
	Variable	CO_2	Energy	GDP	All	Results
CO_2, oil, GDP	CO_2	–	0.37 (0.82)	2.38 (0.30)	2.43 (0.65)	No causality
	Oil	14.06*** (0.00)	–	9.38** (0.05)	21.80*** (0.00)	Directional causality from oil to GDP and CO_2
	GDP	3.62* (0.06)	3.27* (0.09)	–	9.53** (0.04)	Directional causality from GDP to Oil and CO_2
CO_2, gas, GDP	CO_2	–	2.00 (0.15)	1.43 (0.23)	4.63 (0.11)	No Causality
	Gas	4.47** (0.03)	–	4.29** (0.03)	20.45*** (0.00)***	Directional causality from gas to GDP and CO_2
	GDP	8.12*** (0.00)	3.73* (0.09)	–	13.36*** (0.00)	Directional causality from GDP to gas and CO_2
CO_2, coal, GDP	CO_2	–	2.75 (0.16)	3.20 (0.13)	3.64 (0.13)	No causality
	Coal	4.30* (0.08)	–	3.70* (0.09)	9.49*** (0.00)	Directional causality from coal to GDP and CO_2
	GDP	8.54*** (0.00)	5.38* (0.06)	–	14.04*** (0.00)	Directional causality from GDP to coal and CO_2
CO_2, renewable, GDP	CO_2	–	0.51 (0.77)	4.5 (0.11)	5.07 (0.27)	No causality
	Renewable	0.29 (0.86)	–	0.62 (0.73)	4.09 (0.39)	No causality
	GDP	1.09 (0.4)	0.05 (0.97)	–	2.12 (0.25)	No causality

(continued)

Table 4 (continued)

Equation	Independent variables					
	Variable	CO_2	Energy	GDP	All	Results
CO_2, electricity, GDP	CO_2	–	1.42 (0.4)	2.64 (0.22)	0.83 (0.65)	No causality
	Electricity	38.24*** (0.00)	–	3.44* (0.09)	43.02*** (0.00)	Directional causality from electricity to GDP and CO_2
	GDP	12.7*** (0.00)	19.97*** (0.00)	–	18.49*** (0.00)	Directional causality from GDP to electricity and CO_2

***, **, and * Indicate significance at the level of 1%, 5%, and 10%

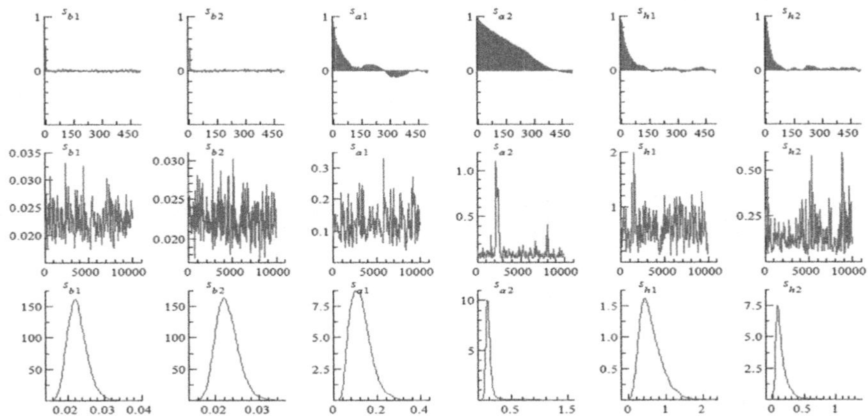

Fig. 2 Results of TVP-VAR model with stochastic volatility for simulated data. The first row (autocorrelation status of samples), the second row (path of the samples), the third row (pre-density of the samples)

gas experienced a constant trend, as the villages and most industries equipped with gas, so we have seen a low size of volatility. According to the Iranian coal and mining company, since past coal consumption has a specific place in Iran's industry sector, which is essential for power generation in power plants. As represented in Fig. 7, the volatility of coal consumption is less than oil and natural gas but slightly similar to natural gas trends. The volatility trend shows a large size during later years when it is related to subsidizing targeting of oil. So, coal consumption in recent years has been significant. For this reason, according to Fig. 7, the stochastic volatilities in coal

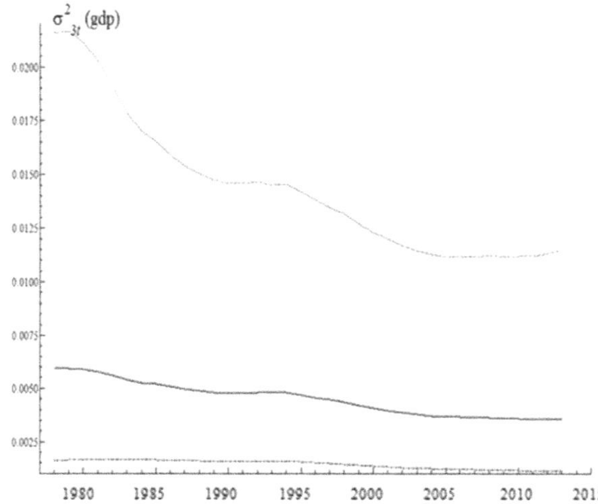

Fig. 3 Stochastic volatility of GDP

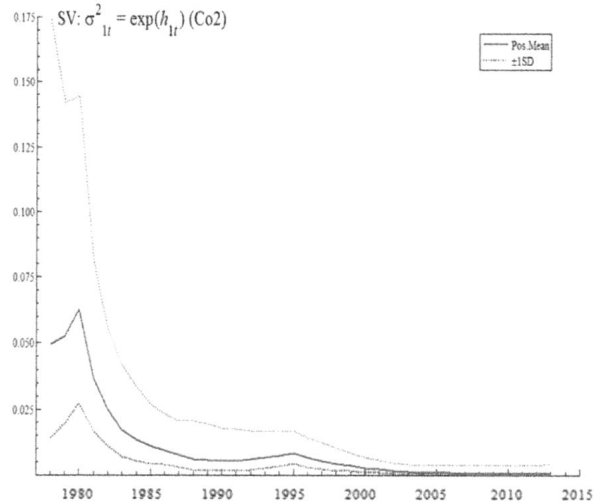

Fig. 4 Stochastic volatility of CO_2

consumption during the high volatility period of oil consumption was less (1986–2001), and increased during 2001–2015. The volatility of electricity consumption was low during the studied period, and it has a constant and slightly decreasing trend due to the moderate use of all sectors and lack of intervention policies. The volatility of electricity consumption shows in Fig. 8.

Response functions (RF) are important tools of VAR methods. The RF examine the response of a variable to the structural shock of the other related variable in the same equation over time. In the TVP-VR method, we can also examine the variable responses to the volatility of variables, which is called "Time-varying impulse

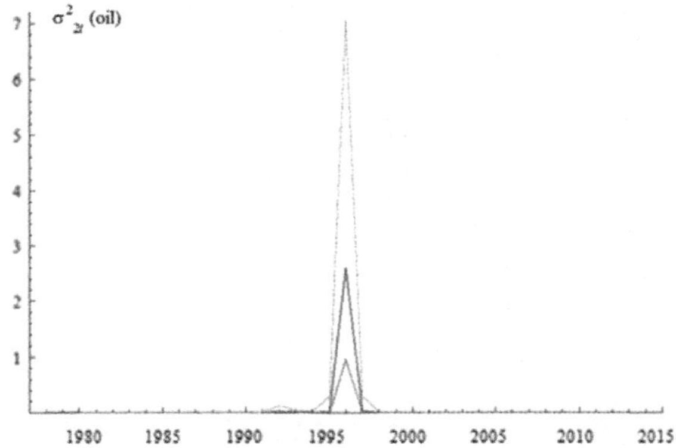

Fig. 5 Stochastic volatility of oil

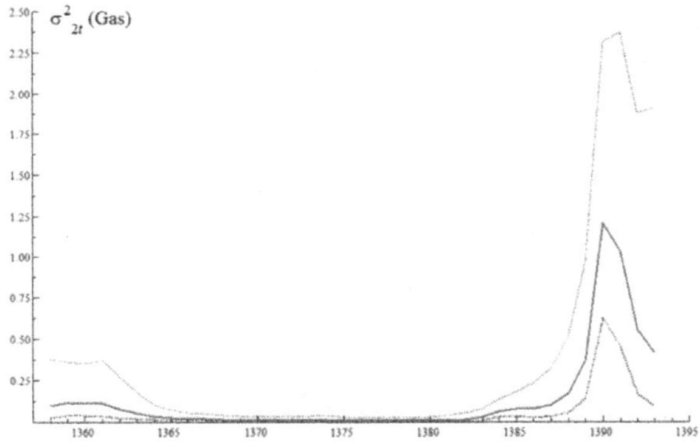

Fig. 6 Stochastic volatility of gas

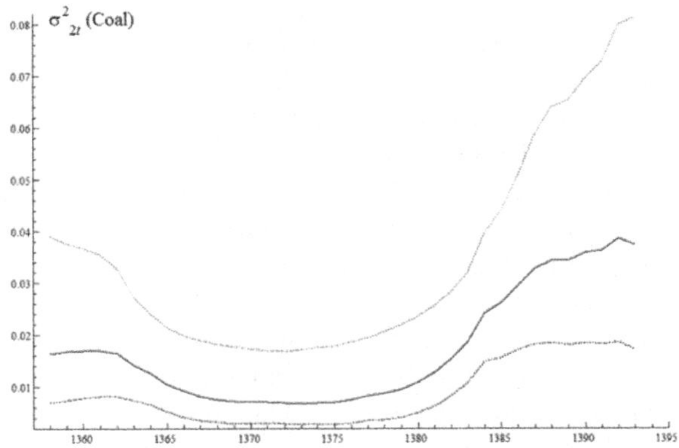

Fig. 7 Stochastic volatility of coal

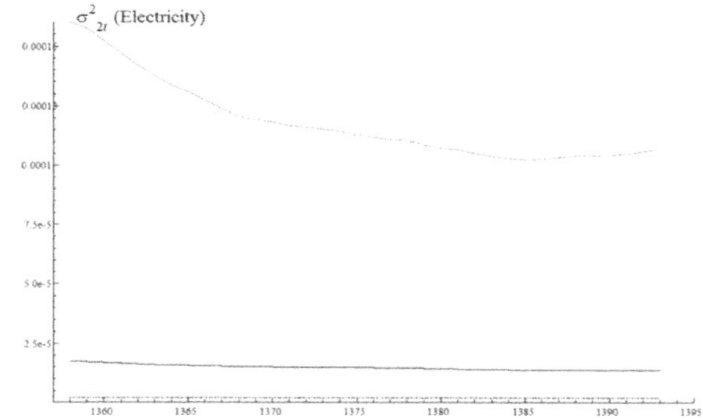

Fig. 8 Stochastic volatility of electricity

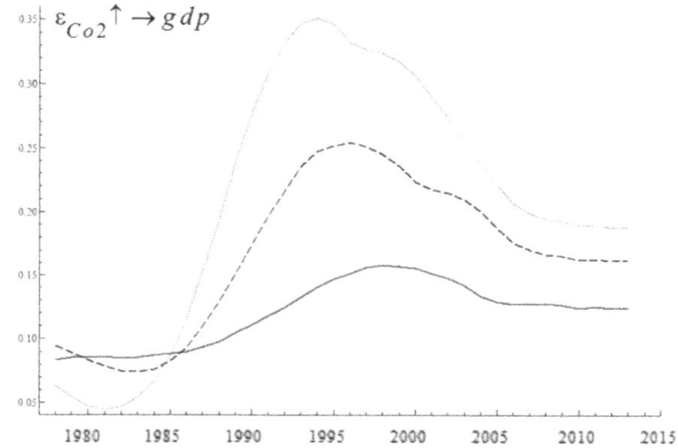

Fig. 9 Response of CO_2 to GDP shocks

analysis." In this study, the reactions figures are drawn with red, purple, and green colors which contribute to the different horizons, next 4, 8, and 12 years. The time-varying response functions and size of responses for the variables are represented in the Figs. 9, 10, 11, 12 and 13.

According to Fig. 9, the response of CO_2 to structural shocks and stochastic volatilities of GDP was positive during the period. From 1978 to 1985, the response was positive and U-shaped after that was positive too but was inverse U-shaped with a peak point in 1997, knows as EKC theory. As GDP increases, CO_2 emission is not mattered but continues by increasing GDP rate environmental problems and environmental-friendly technologies considered to reduce CO_2 emissions. We also find a close and a positive correlation between CO_2 emissions and GDP trends, as that GDP and CO_2 emissions increase and decrease together during the period. Iran's GDP has faced many shocks over the period. Primarily, since 1978–1989 the period coincides with the revolution and the war of Iran–Iraq, then from 1990–1997 was the construction period, and after that, we had different political problems. Earlier in the war, GDP had a fluctuated trend due to fluctuations in OPEC oil prices and exports, but in the construction period, GDP gradually increased despite being embargo and

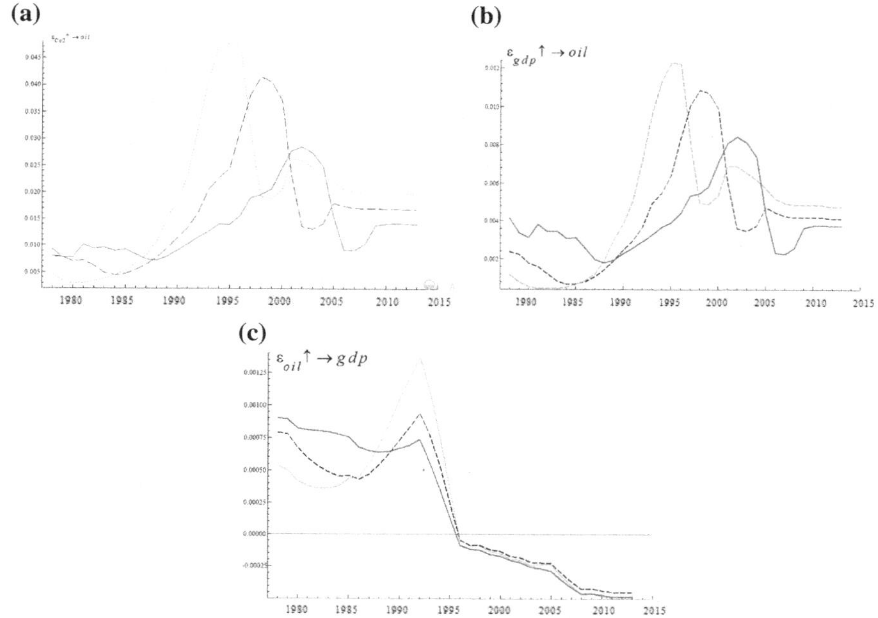

Fig. 10 **a** Response of CO_2 to oil shocks. **b** Response of GDP to oil shocks. **c** Response of oil to GDP shocks

high infrastructure expenditures. According to Pishbahar et al. (2019) from 1997 to 2003, GDP (Iran's foreign exchange revenue) has declined due to the OPEC oil prices. From the beginning of the Iraq–US War in 2003, OPEC oil prices and GDP has increased but declined again due to political problems and embargo since now. Therefore, during the studying period, Iran had many fluctuations and shocks, which have been accompanied by CO_2 emissions. It happened because of OPEC oil prices, which determine the amount of revenue that affects the output of economic sectors. Finally, these products determine the extent of CO_2 emissions. Thus, GDP shocks have an indirect effect on CO_2 emissions. The size of CO_2 response to GDP volatilities is positive all the time and has a large high size of reactions due to the mentioned reasons. It seems that the relationship between GDP volatilities and CO_2 reactions is considerable, and any volatility in GDP leads CO_2 emissions to respond in large size. The response of GDP to CO_2 volatility didn't analyze due to the lack of causality from CO_2 to GDP.

The size and shape of CO_2 response to oil consumption shocks and stochastic volatilities are represented in Fig. 10a. This figure is U-shaped at the beginning of the period and inverse U-shaped in the rest of the years with different peak points, and also CO_2 emission shows a positive and low size of responses to oil volatilities. Oil is one of the main fossil fuel sources in Iran, and one of the main factors of CO_2 emissions, so the size and amount of CO_2 response to oil consumption depend on the how amount consume. As mentioned in the previous section, OPEC oil prices determine

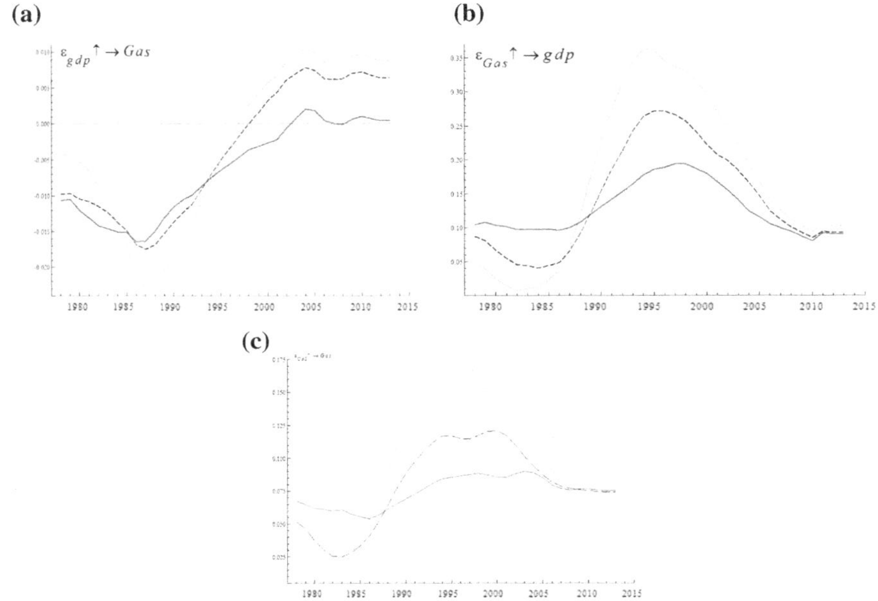

Fig. 11 **a** Response of GDP to gas shocks. **b** Response of gas to GDP shocks. **c** Response of CO_2 to gas shocks

the GDP the amount and also the power of domestic producers to purchasing primary inputs such as oil. Unfortunately, oil prices have experienced a fluctuating trend due to various events, so volatilities and structural shock of oil price indirectly affect oil consumption and also CO_2 emissions. It seems that the CO_2 emission is sensitive to the structural shocks and stochastic volatilities of oil consumptions, which are determined by OPEC oil price. The EKC theory exists between oil consumption and CO_2 emissions.

The response of GDP to oil consumption shocks and stochastic volatilities is as follows. In Fig. 10b, GDP shows the positive and low size of responses to oil shocks and stochastic volatilities, which is U-shaped at the beginning of the period and converted Inverted U-shape in the following. Also, the peak points are not the same in the different horizons of time. From 1990 to 1997, oil consumption and GDP are gradually increased due to the construction period and decreased period of 2009–2015. The response of oil consumption to GDP shocks and stochastic volatilities is different in Fig. 10c; it was positive before 1996 and changed to negative since 1997. From the beginning of the period, oil shows a positive response to GDP with a peak point in 1992 due to the construction period. After 1997 till now, the response of oil to GDP volatilities is negative and sharply decreasing due to embargo and subsidizing targeting policies. By comparing Fig. 10b, c, it seems that the size of GDP response to oil volatilities is more than the size of oil response to GDP volatilities during the period; it is due to that most of the energies in productive sectors are from oil.

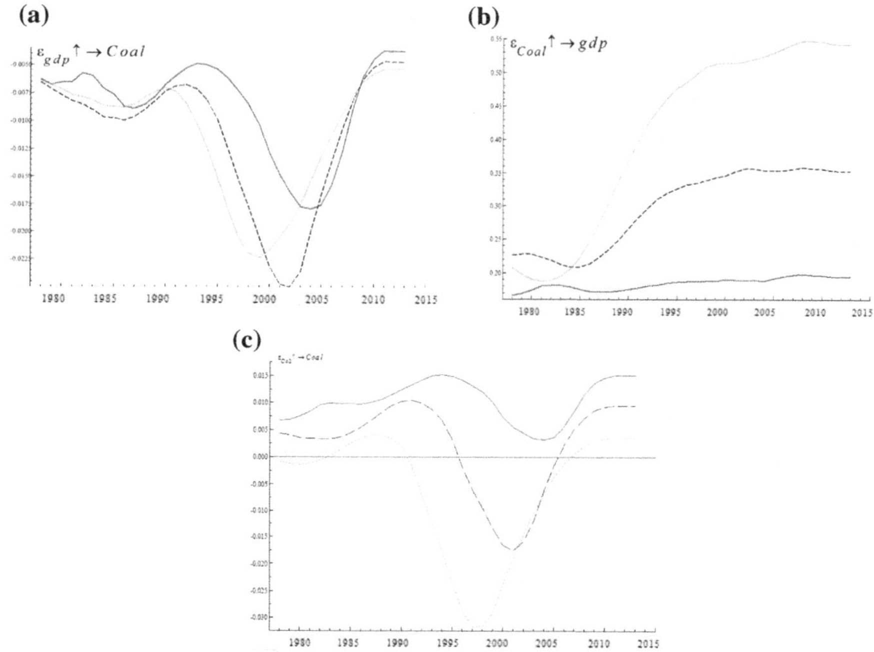

Fig. 12 **a** Response of GDP to coal shocks. **b** Response of coal to GDP shocks. **c** Response of CO_2 to coal shocks

In Fig. 11a, GDP represents a different size of responses to gas volatilities and stochastic shocks, which is negative before 1997 and positive after that. It seems that before 1997, any volatility and shock in GDP caused small and negative responses in gas consumption and positive responses after 1997. Based on statistics before 1997, any decreasing shock in gas consumption associated with an increase in GDP, but it changed after 1997. It seems that this behavior as a low share of gas in the energy-mix of sectors. In Fig. 11b, gas consumption shows a positive and high amount of response to GDP volatilities and structural shocks during the period. It was U-shaped in the beginning and converted inverted U-shape in the rest of the years. Gas consumption volatilities and structural shocks had a considerable effect on GDP, which relates to the dependency of sectors to fossil renewable in the production process.

CO_2 represents a large size of the response to gas consumption shocks and volatilities in Fig. 11c. It means that any volatility in gas consumption led the CO_2 emission to shows high responses; also it is clear that EKC theory exists in this case. As gas consumption increases, CO_2 emission wasn't mattered, but after 1997 by increasing gas consumption, environmental problems, and environmental-friendly technologies reduced CO_2 emissions.

The response of GDP to coal consumption volatility and structural shocks represents in Fig. 12a. The response was mainly U-shaped, negative, and low size, but it

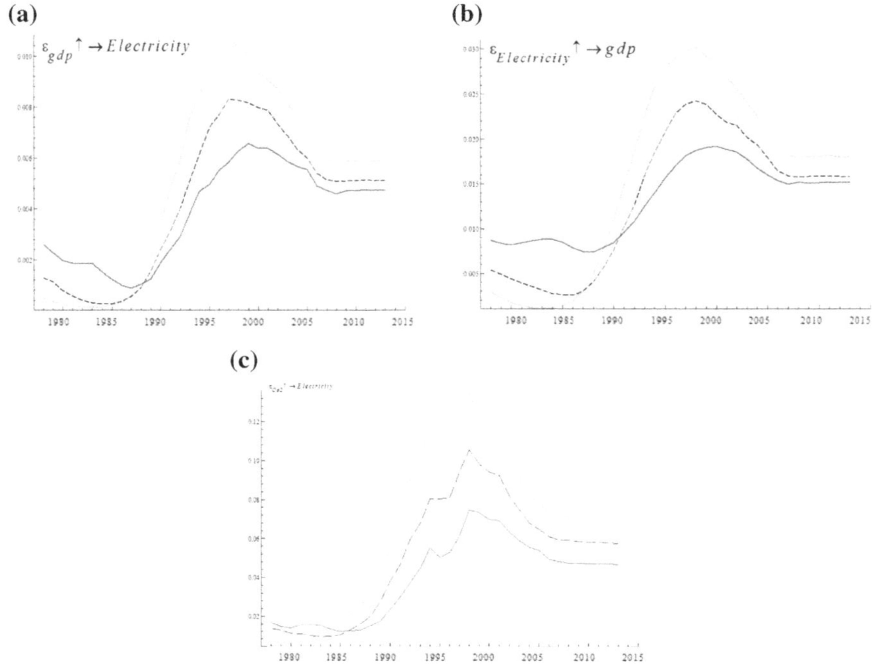

Fig. 13 **a** Response of GDP to electricity shocks. **b** Response of electricity to GDP shocks. **c** Response of CO_2 to electricity shocks

is increasing in the last years. It shows us any volatility and structural shock in coal consumption, led the GDP to respond in a negative way because coal has a small share (under 2%) in the energy-mix of sectors. In Fig. 12b, coal consumption shows an increasing trend of responses to GDP volatility and structural shocks, and also the size of responses is considerably during the period. It means that any volatility and structural shock in coal consumption led the GDP to respond in large size due to that coal is one of the energy sources in economic sectors, especially industry. CO_2 response to coal volatilities and structural shocks represent in Fig. 12c. The responses trend is unclear, which is negative or positive in different years. Based on Fig. 12c, any volatility and structural shocks in coal consumption, led CO_2 emissions to respond in a negative way and small size. The notable point is that coal consumption and CO_2 emissions do not follow the Kuznets' theory, while GDP, oil, and gas consumption follows the Kuznets' theory.

Figure 13a–c represent the response of GDP, electricity consumption, and CO_2 to structural shocks and volatility of electricity consumption and GDP, which had a low amount of response size during the period. Based on Fig. 13a, from 1978 to 1990, the response was U-shaped and then was inverse U-shaped with a peak point in 1997. Similarly, in Fig. 13b, the response of electricity consumption to GDP volatilities and structural shock was U-shaped during 1978–1990 and was inverse U-shaped with a peak point in 1997. By comparing both of them, it seems that the size of electricity consumption volatility and structural shocks response to GDP was more than the response of GDP to electricity consumption, so electricity consumption is sensitive to GDP volatilities and structural shocks. Figure 13c shows the response of CO_2 to electricity consumption volatility and structural shocks, which is positive during the period. From 1978 to 1990 the response was U-shaped and then was inverse U-shaped with a peak point in 1997, and it knows as EKC theory. As electricity consumption increases, CO_2 emission wasn't mattered, but after 1997, by increasing electricity consumption, environmental problems and environmental-friendly technologies considered to reduce CO_2 emissions. It seems that the relationship between electricity volatilities and CO_2 reactions is considerable, and any volatility in electricity consumption leads to CO_2 emissions to respond in large size.

5 Conclusions

Empirical evidence shows that there is a relationship between economic growth and environmental degradation; also the level of economic development is one of the factors in environmental problems. Economic growth is considered as one of the most important goals as well as in Iran. But unfortunately, after Iran's revolution, GDP has experienced significant fluctuations and considerable structural shocks from 1978 till now. The CO_2 emissions due to high demand and consumption of fossil fuels highlight the fluctuating and sharply increasing trend in the last decades. This study investigates the relationship between economic growth, consumption of energy types (oil, gas, coal, renewable, and electricity) and CO_2 emissions by using the TVP-VAR method during 1978–2015. Primary, the stationary of variables, number of co-integration vectors, and causality among the variables are determined by related tests. The results indicate that there is a bidirectional relationship between growth and energy consumption, but was a directional relationship from GDP to CO_2 emissions as well as GDP to a different type of energy consumptions, except for renewable. Investigating the U-shape response functions of the GDP and CO_2, knows as the Kuznets' hypothesis, is dynamically observed in Iran during the studying period. The results of this study are consistent with the results of Mezghani and Haddad (2016), who examined dynamic Kuznets' hypothesis for income and CO_2 emissions in Saudi Arabia. According to Fetras and Nasrindoost (2009), Fetras et al. (2010), Moghaddasi and Rahimi (2010), Hosseininasab and Paykari (2012), Mahdavi Adeli and Nazari (2015), Alizadeh and Golkhandan (2016), Salmanpour et al. (2017) and Airkhizi et al. (2018), whose approved the Kuznets' hypothesis in Iran They didn't pay attention to dynamic form of relationship, stochastic volatilities, and structural shocks of the variables. As we know, most of the variables due to unpredictable events can experience stochastic volatilities or structural shocks, so considering these features can lead to more accurate analysis in policymaking. Applying the TVP-VAR model, which can consider the time-varying impulse analysis, the effect of volatilities, and structural shocks at the different horizon of time, let us decide accurately. Therefore, the results of this study could be a serious warning to policymakers.

The results of the TVP-VAR model are in three main groups. First, the size of the response of energy consumption and CO_2 emissions to GDP volatilities and structural shocks are as follow: coal, CO_2 emissions–natural gas–electricity and oil. Second, the size of the response of CO_2 emissions to volatilities and structural shocks of energy consumption are: oil–gas–electricity and coal. Third, the size of the response of GDP to volatilities and structural shocks of energy consumption are: oil–gas–electricity and coal. Also, the size of responses in the first group is more than others; it means that GDP sensitivity to volatility and structural shocks of energy consumption is less than the sensitivity of energy consumption and carbon to GDP. Because most of Iran's revenue is provided by oil exports, it is not mainly dependent on non-oil exports. On the other hand, the large size of respond of CO_2 to GDP is considerable; this is due to the high share of fossil fuels in the energy mix of sectors. It is necessary that policymakers pay attention to high volatility in GDP to avoid its effects on energy consumption and CO_2 emissions, which is possible by replacing renewable energies such as solar and wind in the economy.

References

Airkhizi MRN, Salmanpoor Zoonooz A, Shokouhifard S (2018) Theorical and empirical analysis of Kuznets environmental pollution in the Iran during 1986-2016. Environ Cross sectional Devel 3:29–46

Alam MJ, Begum IA, Buysee J, Sanzidur R, Huylenbroeck GV (2011) Dynamic modeling of casual relationship between energy consumption, CO_2 emissions and economic growth in India. Renew Sustain Energy Rev 15:3243–3251

Alizadeh M, Golkhandan A (2016) Energy consumption and economic growth in OPEC countries: new empirical evidence of panel integration and interdependence. Quart J Policy Plan Energy Plan 2(5):131–164

Barimani M, Kaabinezhad A (2014) Renewable energies and sustainable in Iran. Scienti Pap Renew Renew Energ Year 1(1)

Barnett A, Mumtaz H, Theodoridis K (2014) Forecasting UK GDP growth and inflation under structural change: a comparison of models with time-varying parameters. Int J Forecast 30:129–143

Chindho S, Abdurahim A, Waziri S, Houng W, Ahmad A (2015) Energy consumption, CO_2 emission and GDP in Nigeria. GeoJournal 80:315–322

Dinda S (2004) Survey environmental Kuznets curve hypothesis: a survey. Ecol Econ 49:431–455

EIA (2015) U.S. Energy information administration. Overview 2015

Environmental Assessment Agency (2016) Energy and climate change

Fetras MH, Ghaffari H, Shahbazi A (2010) Study of the relationship between air pollution and economic growth of oil exporting countries. J Econ Growth Dev 1(1):78–60

Fetras MH, Nasrindoost M (2009) Investigating the relationship between air pollution, water pollution, energy consumption and economic growth in Iran. Quart J Energy Econ 6(21):135–113

Fodha M, Zaghdoud O (2010) Economic growth and pollutant emission in Tunsania: an empirical analysis of the environmental Kuznets curve. Energy Policy 38:1150–1156

Hosseininasab A, Paykari S (2011) Investigating the impact of economic growth and commercial liberalization on environmental pollution. J Econ Issu Policy Rev 9:82–61

Kuznets S (1955) Economic growth and income inequality. Am Econ Rev 45:1–28
Lee CC (2006) The causality relationship between energy consumption and GDP in G-11 countries revisited. Energy Policy 34:1086–1093
Lim KM, Lim SL, Yoo SH (2014) Oil consumption, CO_2 emission, and economic growth: evidence from the Philippines. Sustainability 6:967–979
Mahdavi A, Amirbabaei S (2013) The effect of financial development on environmental quality in Iran. Quart J Econ Res 15(4):23–1
Mahdavi Adeli A, Nazari R (2015) Economic growth, energy and environment: E3 model in Iran. Quart J Econ 11(1):40–19
Mezghani I, Haddad HB (2016) Energy consumption and economic growth: an empirical study of the electricity consumption in Saudi Arabia. J Renew Sustain Energy Rev 75:145–176
Moghaddasi R, Rahimi R (2010) Investigating the relationship between air pollution and economic growth in selected provinces of Iran: application of Kuznets environmental curve. Quart J Econ Sci 3(11):37–19
Ministry of Energy (2018) Office of planning for electricity and energy
Nazari R, Adeli Mahdavi MH, Dadgar Y (2014) Investigating factors affecting pollution in Iran during the period of 1974–2014. J Econ Growth Develop 6(21):60–47
Nakajima K (2011) Time-varying parameter VAR Model with stochastic volatility: an overview of methodology and empirical applications. IMES discussion paper Japan
Oganesyan M (2017) Carbon emission, energy consumption and economic growth in the BRIC. Master Thesis in Economic Jonkoping University Sweden
Ozturk I, Acaravci A (2010) CO_2 emission, energy consumption and economic growth in Turkey. Renew Sustain Energy Rev 14:3220–3225
Primiceri GE (2004) Time varying structural vector auto regressions and monetary policy. Working Paper Princeton University
Pakrooh P, Pishbahar E (2019) Forecasting air pollution concentration in Iran, Using a hybrid model. Pollution 5(4):739–747
Pishbahar E, Pakrooh P, Ghahremanzadeh M (2019) Effects of oil prices and exchange rates on imported inputs price for the livestock and poultry industry In: Rashidghalam M (ed) Sustainable agriculture and agribusiness in Iran. Perspectives on development in the Middle East and North Africa (MENA) Region. Springer, Singapore
Salmanpour A, Mousavi SK, Shokohifard S (2017) Effect of economic growth, energy consumption and financial development on environmental pollution in Iran during the period of 1986–2016. Environ Sci Stud 2(1):111–120
Sharifi A, Kiani GH, Khoshakhlag R, Bagheri MM (2012) Assessment of the replacement of renewable energies in place of fossil fuels in Iran: an optimal control approach. Econ Model Res J 11:123–140
Tekaleg E, Mulugeta W, Hussen M (2016) Energy consumption, carbon dioxide emission and economic growth in Ethopia. Glo J Manage Busin Res 16:1–10
Wolde ET, Mulugeta W, Hussen MM (2016) Energy consumption, carbon dioxide emissions and economic growth in Ethiopia. Glo J Manage Busin Res 16(2):1–9
Zhang XP, Cheng XM (2009) Analysis energy consumption, carbon emission and economic growth in china. Ecol Econ 68:2706–2712

Parisa Pakrooh is a Ph.D. candidate at the University of Tabriz, where she furthers her research on Natural Resource Economics and Game Theory. She has a Bachelor's degree is in Agricultural Economics from University of Tabriz. She has a Master's degree in Agricultural Policy from University of Tabriz. She has been researching Agricultural Policy and Natural Resource Economics.

Esmaeil Pishbahar is Associate Professor of Agricultural Economics at University of Tabriz, Iran. He holds a B.Sc. in Agricultural Economics from University of Tabriz and a M.Sc. in Agricultural Economics from University of Tehran. He did his Ph.D. in Science Economics at departments of Economics and Management, University of Rennes 1, France. His areas of interest and research are Applied Econometrics, Agricultural Risk Management and Insurance, and International Trade. His teaching area are Advanced Econometrics, Mathematical Economics, and Macroeconomics at under- and postgraduate levels. He has over 100 publications in journals and chapters in books.

Oil Price Volatility and Food Price Linkage: Evidence of Dutch Disease in Iran's Agricultural Sector

Zahra Rasouli, Mohammad Ghahremanzadeh, and Masoomeh Rashidghalam

Abstract This paper seeks to establish the relationship between oil price volatility and domestic food price inflation in Iran. Different generalized autoregressive conditional heteroskedasticity (GARCH)-type models are estimated to model oil price volatility. Based on the multiple loss functions, periodic GARCH (PGARCH) model is selected as the best. The estimated volatility, together with nominal exchange rate and basic food price inflation are included in a VECM model to estimate the co-integrating vector. The findings reveal that there are a positive and highly significant relation between food price inflation and oil price volatility and also a negative and significant relation between food price inflation and exchange rate. This long-run relation proves the existence of Dutch disease in Iran's economy.

Keywords Dutch disease · Price volatility · Inflation · Vector error correction model · Autoregressive conditional heteroscedasticity · Food · Oil · Exchange rate

1 Introduction

There has been increasing attention to the relationship between food and energy prices in the world recently. A combination of high oil and food prices, due to its potential effects on growth, inflation, and income distribution has been become a decreasing element of stability in the global economy. In terms of the effects of these two factors on income distribution, poverty, and inflation, food prices are more important than energy prices. Uncertainty, distortion, and gradual decline in purchasing power make

Z. Rasouli · M. Ghahremanzadeh (✉)
Department of Agricultural Economics, University of Tabriz, Tabriz, Iran
e-mail: ghahremanzadeh@tabrizu.ac.ir

Z. Rasouli
e-mail: zahrarasouli.480@gmail.com

M. Rashidghalam
Centre for Entrepreneurship and Spatial Economics (CEnSE), Jönköping International Business School (JIBS), Jönköping, Sweden
e-mail: masoomeh.rashidghalam@ju.se; maso.azar@gmail.com

M. Rashidghalam (ed.), *The Economics of Agriculture and Natural Resources*, Perspectives on Development in the Middle East and North Africa (MENA) Region,
https://doi.org/10.1007/978-981-15-5250-2_11

it necessary to take the measures at the national and international levels to get rid of the unfavorable economic situation (Von Braun 2007). Oil and food prices are highly correlated. For example, in Iran the correlation between consumer price index and light crude oil price is about 0.79. However, the task of crafting appropriate policy responses to the food crisis is made harder by rising oil prices and ensuing fiscal and balance of payments problems.

Agricultural production in Iran has been quite erratic. Figure 1, part (a) shows the total agricultural production, including crops, livestock, and fisheries during the period 1968–2011. Despite the relative increase in production over the years, in most of the times production loss is recognized. Average growth rate of production was 1.64% and the maximum drop of −14.69% was in 2009. Part (b) of this figure indicates that agricultural subsidy experienced mild increase until 2004 and had a mutation in 2005 and 2006. It seems that from 2004 onward, agricultural subsidies have been faced with a flurry. For example, it had for about 29% increase in 2006, and it dropped to 11%, in 2010. Part (c) which shows total credit disbursed by the Agricultural Bank to this sector, had a low growth of 2% in 2007 and 2008 and also it was about −4% in 2010. According to parts (a) and (d) in Fig. 1, we had greater amount of import during the years that production decline was greater. So that the

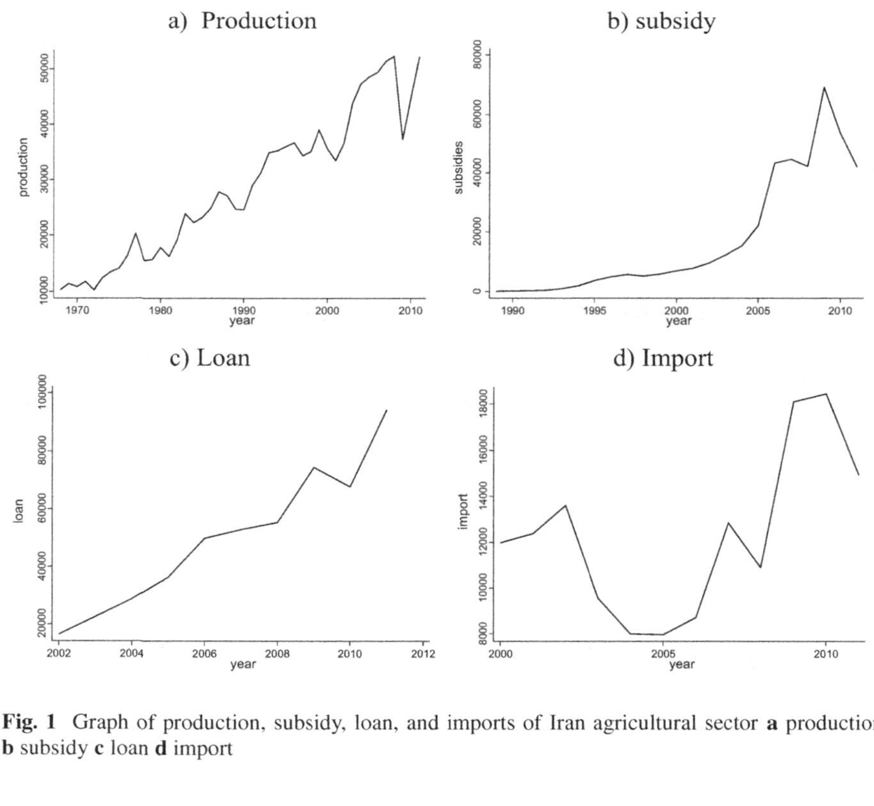

Fig. 1 Graph of production, subsidy, loan, and imports of Iran agricultural sector **a** production **b** subsidy **c** loan **d** import

highest rate of growth in agricultural imports (22%) was in 2009, which the largest drop in production was observed. Also, Imports lowest level occurred in 2003–2006.

Iran's oil and natural gas reserves are among the world's largest, and its economy depends significantly on the extraction of these resources (Rashidghalam 2019). The relationship between the oil sector and food production in Iran can be considered as a typical example of the "Dutch Disease Syndrome". In general, Dutch Disease describes a reduction in a country's export performance as a result of an appreciation of the exchange rate after a natural resource has been discovered. The increase in revenue from the natural resources hurts traditional exports or tradable sectors (such as local manufacturing and agricultural exports) through an increase in the exchange rate. Additional government revenue also implies greater government expenditure and movement of resources, like labor and capital from the non-booming tradable sectors (like agricultural export sector) to the booming sector (like oil sector) and government sector. If these resources are scare (as usually the case) then the prices of goods and services which are produced by employment of these resources will rise in response to the higher demand. The base model of Dutch Disease Syndrome that presented in Corden and Neary (1982), Corden (1984), assumes three sectors; the booming sector, the lagging sector and the non-tradeable sector. The first two sectors produce goods which are traded in the international market at specified world prices. In this model it is assumed that output in each sector is produced using a sector-specific factor and labor, which is transferable between all three sectors. All factor prices are flexible and all factors are internationally non-transferable.

An exogenous increase in the price of one of the tradable sectors' output or a windfall discovery of new resources, make a boom in that sector. As a result of the boom, the aggregate income of the factor initially employed in that sector will increase. The main characteristic of the analysis is a distinction between two effects of the boom, namely the resource movement effect and the expenditure effect. When the marginal product of labor in booming sector increases, the demand for labor will rise in this sector. In a country like Iran, where the booming sector is unable to fully absorb the labor from the lagging and non-tradable sectors, it will result in severe structural dislocation and underemployment. Therefore, we assume that lagging sector represents the agricultural exports sector and the booming sector is the oil sector, while the non-tradable sector is the basic food products sector. Due to boom in the oil sector, mentioned spending effect, will shift and raise basic food crops demand, thus will raise the price of this crops. Increase in food prices will lead to transfer of factors from agricultural exports sector into basic food sector. However, it is assumed that the level of resources drawn from the food production sector as a result of the boom in the oil sector is larger than the resources attracted into it, hence supply will decrease. Therefore, at the equilibrium basic food prices will be much higher (Udoh and Egwaikhide 2012; Eifert et al. 2002).

A sizable literature on the Dutch Disease can be found in the economics literature. For example, Eltony (2002) using a vector autoregressive model (VAR), investigates the impact of oil price fluctuations on macroeconomic variables of the Kuwaiti economy. The empirical results highlight the causality running from the oil prices, to macroeconomic variables. Hamilton (2003) provides a flexible approach to characterize the nonlinear relation between oil price changes and GDP growth. His paper reports clear evidence of nonlinearity relation between oil price and government activities. So that, an increase in oil price has a significant effect on production decrease in industrial countries while decreasing oil price doesn't have important effect on their economy. Puyana (2000) assesses the effects of oil revenues on the Colombian agricultural sector for the period of 1980–1994. Results indicate that in Colombian economy especially in agricultural sector Dutch Disease exists. This means that oil revenue increase moves agricultural employee to other economic sectors. As a result, the growth rate of agricultural production, cultivation and labor productivity have been declined sharply. Lartey (2006) investigated the relationship between real exchange rate and capital flows with emphasis on foreign direct investment in 16 countries of Saharan Africa for the period 2000–1980. In his model, the real exchange rate is a function of foreign direct investment, private investment, foreign aid, government spending, excess money growth, and the degree of openness of economy. The result showed that the increase in foreign direct investment and foreign aid strengthens the real exchange rate. Therefore, Dutch disease was observed in these 16 African countries. Udoh and Egwaikhide (2012) investigate the causality relationship between domestic food price inflation and international oil price fluctuations in Nigeria during 1970–2008. The results of this study show that there is a causal relationship between these two variables. The Granger causality test indicates that causality runs from international oil price to domestic food price. Positive relationship among domestic food price inflation and oil price fluctuations is suggestive of the existence of "Dutch Disease" in this economy. Another study by Nwoko et al. (2016) examine the long- and short-run relationships between oil price and food price volatility as well as the causal link between them in Nigeria. In this study, the Johansen and Jesulius co-integration test revealed that there is a long-run relationship between oil price and domestic food price volatility. Bakhtiari and Haggi (2001) studied the impact of increase in oil revenues on agricultural sector in Iran's economy for the period of 1961–98. Results show that the Dutch disease in Iran's economy has emerged as an anti-agricultural phenomenon. This means that as oil revenues rises, employment transfers from agriculture to service and industry sectors. Therefore, it causes a decrease in the share of agricultural value added and increase in relative prices of service and industry sectors. Pasban (2004) analyzed the effects of oil price volatility on agricultural production using data for the period of 1971–2000. In this study the hypothesis, that the oil boom reduces Iran's agricultural production, has been tested. The results showed that this hypothesis is confirmed for the economy and oil prices have negative impact the agricultural sector. On the other hand, effects of oil price shocks on the agricultural sector have declined over time and disappeared.

The aim here is to investigate the relationship between oil price volatility and food price. The paper is organized as follows: Sect. 2 presents Materials and methods; Sect. 3 describes the empirical results and finally conclusions are provided in Sect. 4.

2 Methodology

This paper seeks to establish the relationship between oil price volatility and domestic food prices. Price volatility is not observable and must be estimated. We use GARCH-type models for modeling and estimating oil price volatility. Different linear and nonlinear models are compared based on seven loss functions defined in the study of Hansen and Lunde (2001), as MSE_2, MSE_1, PSE, QLIKE, R2LOG, MAD_2, MAD_1. Finally, the model of PGARCH that was suggested by Higgins and Bera (1992) is selected. This model is defined as:

$$h_t^{\delta} = \omega_0 + \sum_{i=1}^{p} \alpha_i(\varepsilon_{t-i}^{\delta}) + \sum_{i=1}^{q} \beta_i(h_{t-i}^{\delta}) \tag{1}$$

where h is the conditional variance and ε is the residual of conditional mean model. α, β, and ω are the parameters of the model and are estimated by maximum likelihood. Volatility estimates from the PGARCH model with exchange rates and food price inflation are used in a co-integration model to investigate the presence of long-run relationship between oil price volatility, food price inflation, and exchange rate. Johansen (1988) and Johansen and Juselius (1990) introduced linear multivariate co-integration test based on unconstrained vector autoregressive model (VAR). We use the vector error correction model (VECM):

$$\Delta P_t = \Pi P_{t-1} + \sum_{i=1}^{k-1} \Gamma_i \Delta P_{t-i} + \Phi D_t + \varepsilon_t \tag{2}$$

where P_t includes n variable and Π, Γ_i, Φ are matrix of coefficients to be estimated. D_t is a vector of deterministic variables (such as constant terms, trends, and dummy variables) and ε_t is a vector of error terms which is assumed to be white noise. The focus of Johansen's method is on the matrix Π and its rank. The rank of this matrix (r) represents the number of stable linear combinations in P_t or in the other words represents the number of co-integrated relationships in the system. When r is equal to zero, there are no co-integrating relations and none of the linear combinations are stable. Therefore, the model is just a VAR in differenced data. At the other extreme, if $r = k$ then the original price series is most likely already stationary. However, the case of interest is when $0 < r < k$; then there exists long run relationship among some of the variables, hence there is r co-integrating vector or r stable linear combination (Niquidet and Manely 2008). To determine the rank of matrix Π, trace test and the maximum eigenvalue test can be used.

Dataset consists of monthly exchange rate, price of light crude oil, and food price inflation. We utilize secondary data from the Central Bank of Iran covering the period April 2002–November 2012. Food price inflation (FINF) was calculated as percentage variation of food price.

3 Results and Discussion

As mentioned, this paper tries to investigate the relationship between oil price volatility and food price inflation. The methodology adopted in this study includes estimation of oil price volatility, test for stationarity of variables, co-integration test, determining co-integration rank, estimating VECM model, and finding co-integrating vectors. The results will be presented in detail.

The right and left parts of Fig. 2 present consumer price index and Food price inflation (FINF), respectively, for the period of 2002–2012. The graph of Food Price Index indicates that this index increased gradually during 2002–2008, and accelerated subsequently. This indicates that Iran's major agricultural products were affected from the 2007 and 2008 global food price crisis. In the left figure, food price inflation (FINF), represents relatively large fluctuations in food prices during the study period.

To obtain the price volatility, we used monthly data of Iran light oil price between 2002 and 2012, and adopted GARCH-type models. Table 1 compares different models based on MSE_2, MSE_1, PSE, QLIKE, R2LOG, MAD_2, MAD_1. It seems that MA(6)/PGARCH(1,1) model is relatively best performing model. Estimates of the residuals and conditional volatility of this model are plotted in Figs. 3. Oil price volatility is estimated as the conditional standard deviation.

After estimating oil price volatility, the next step is to test for stationarity of the variables. Tests results are presented in Table 2. They reveal that the null hypothesis of a unit root is accepted in all three time series.

The results of unit root tests reveal that the null of a unit root test is accepted for the level series. Therefore, we conclude that the series are stationary. Based on this, co-integration test is done. Results of Johansen co-integration test are summarized in Table 3. According to this table, the rank of co-integration matrix is one, this implies that only one long-run relationship between three variables exist. To obtain the co-integrating vector, a VECM model with 12 lags was estimated. Based on the

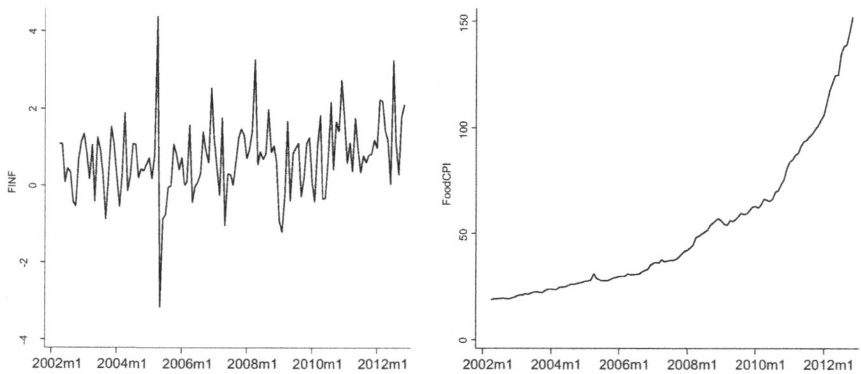

Fig. 2 Graphs of food price index and FINF in Iran, 2002–2012

Table 1 Comparison between different estimated volatility models for oil prices

	MSE_2	MSE_1	PSE	QLIKE	R2LOG	MAD_2	MAD_1
AR(5)/GARCH(1,1)	0.3952	9.1365	111876.1953	166.7107	19.3065	0.0082	0.0680
MA(5)/GARCH(1,1)	0.3873	9.0961	103330.6250	162.5557	18.9249	0.0082	0.0681
MA(6)/GARCH(1,1)	0.3342	8.0431	122439.1719	167.8883	**18.5838**	0.0074	**0.0626**
AR(6)/EGARCH(1,1)	0.2667	7.7701	98547.1797	172.7125	19.4309	0.0073	0.0643
MA(6)/EGARCH(1,1)	0.2617	7.7362	90344.4297	167.1389	19.5915	0.0073	0.0652
TGARCH(1,1)	0.3484	8.8630	87879.7031	164.5401	20.3740	0.0081	0.0685
AR(12)/TGARCH(1,1)	0.3654	8.9505	102844.0625	171.7979	21.3302	0.0081	0.0684
MA(1)/TGARCH(1,1)	0.2753	8.3392	**86103.2891**	164.5684	24.8943	0.0077	0.0679
MA(12)/TGARCH(1,1)	0.3697	9.0066	105906.8359	172.5570	20.0447	0.0082	0.0684
ARMA(5,5)/TGARCH(1,1)	0.3728	8.8241	150098.2344	183.9212	19.9224	0.0079	0.0661
GJR(1,1)	0.3334	9.0764	87028.5859	161.8422	19.4692	0.0080	0.0687
AR(9)\GJR(1,1)	0.3187	8.8349	100259.3828	174.1380	20.0777	0.0077	0.0671
MA(9)\GJR(1,1)	0.3133	8.7683	99384.7500	173.3788	19.8306	0.0077	0.0670
SAGARCH(1,1)	0.3409	8.7903	87024.5313	**161.7085**	19.4830	0.0080	0.0683
AR(8)\SAGARCH(1,1)	0.3159	8.3913	95519.7734	167.9947	19.7199	0.0078	0.0665
MA(12)\SAGARCH(1,1)	0.3673	8.9365	102776.7969	166.9858	19.8748	0.0081	0.0683
ARMA(5,5)\SAGARCH(1,1)	0.3624	8.7637	129714.8984	180.0577	20.4486	0.0079	0.0660
MA(6)\PGARCH(1,1)	**0.2375**	**7.4207**	108096.4453	165.4724	19.9999	**0.0070**	0.0631
NGARCH(1,1)	0.3399	8.7941	88410.3125	162.6685	19.4805	0.0080	0.0683
AR(8)\NGARCH(1,1)	0.3150	8.3926	97215.9063	169.0861	19.7595	0.0078	0.0665
MA(12)\NGARCH(1,1)	0.3663	8.9426	104806.9375	167.9842	19.9530	0.0081	0.0683

(continued)

Table 1 (continued)

	MSE_2	MSE_1	PSE	QLIKE	R2LOG	MAD_2	MAD_1
ARMA(5,5)\NGARCH(1,1)	0.3610	8.7714	132084.5781	181.1755	19.6951	0.0079	0.0660
NGARCHK(1,1)	0.3399	8.7941	88410.3125	162.6685	19.4805	0.0080	0.0683
AR(8)\NGARCHK(1,1)	0.3150	8.3926	97215.9063	169.0861	19.7595	0.0078	0.0665
MA(12)\NGARCHK(1,1)	0.3663	8.9426	104806.9375	167.9842	19.9530	0.0081	0.0683
ARMA(5,5)\NGARCHK(1,1)	0.3610	8.7714	132084.5781	181.1755	19.6951	0.0079	0.0660
AR(1)\APGARCH(1,1)	0.3537	8.9073	89214.3281	163.0616	19.4417	0.0081	0.0684
MA(3)\NPGARCH(1,1)	0.3729	8.7796	92622.6406	169.0378	19.9487	0.0081	0.0681
ARMA(5,6)\NPGARCH(1,1)	0.3143	8.1418	107727.3594	183.4928	19.8822	0.0074	0.0644

Note The formula of these loss functions are presented in Hansen and Lunde (2001, p. 10)
The value of MSE_1 and MSE_2 are product by 1000

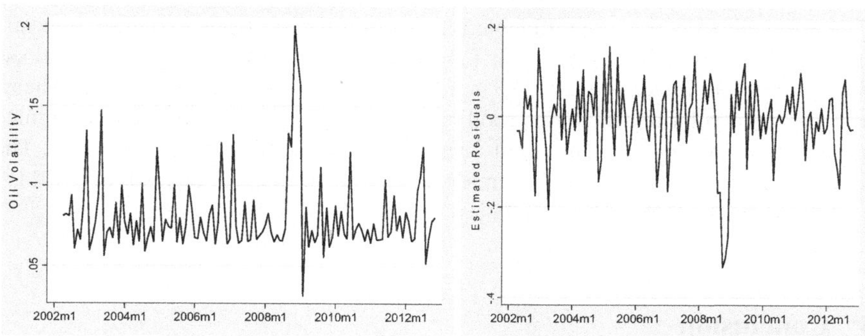

Fig. 3 Plots of residuals (ε) and conditional volatility (h) estimated from MA(6)/PGARCH(1,1) model

Table 2 Unit root test

Nominal exchange rate		Oil volatility		Food inflation		
First differences	Level	First differences	Level	First differences	Level	
5.712	−2.130	−4.675	−3.507	−4.198	−1.671	DF-GLS
5.002	−0.795	−11.208	−1.999	−4.774	−0.073	ADF
0.827	0.209	0.00909	0.0731	0.0697	0.598	KPSS

Table 3 Johansen co-integration test

1% critical value	5% critical value	Trace statistic	Maximum rank
29.75	24.31	32.2662	0
16.31	12.53	10.6466	1
6.51	3.84	0.1010	2
		Max statistics	
22.99	17.89	21.6195	0
15.69	11.44	10.5456	1
6.51	3.84	0.1010	2

estimated Π matrix the co-integrating vector is:

$$FINF = 124.48 * OilVolatility - 0.0011 * NominalExchangeRate$$
$$(0.000)^{*} \qquad (0.000)^{*}$$

The coefficient of oil price volatility was positive and strongly significant statistically (124.48) which provided evidence in support of Dutch Disease phenomenon.

This means that the booming tradable sector (oil industry) leads to higher domestic price in the non-tradable sector (basic food sector). The results show that 1% increase in oil price volatility leads to 124.5% increase in the basic food price inflation. Therefore, oil price volatility increases food price. Increased oil revenues could stagnate in the national economy. This event occurs due to reduction in exchange rate. The results show that one percentage decrease in nominal exchange rate leads to 0.0011% increase in the basic food price inflation.

4 Conclusion

The aim of this study was to investigate the impact of oil price volatility on food price inflation in Iran. However, among the oil-exporting developing countries, rising oil prices create windfalls which lead to increase in government revenues and income. Mismanagement of such windfalls and the ensuing corruption in the polity heat up the economy and probably lead to worst economic outcomes including inflation. Based on the empirical analysis and finding in this paper, it is obvious that Dutch Disease does exist in Iran's economy. Oil price volatility has a long run relationship with food price inflation in Iran. So that, Iran's oil industry in 2002–12 increased food prices in the country. Therefore, the volatility of oil price will increase the fluctuation in food prices and will cause more inflation.

References

Bakhtiari S, Haggi Z (2001) Effects of rising oil revenues on agriculture: the case of Dutch disease in Iran. Agric Econ Dev 35:109–139 (In Farsi)

Corden WM (1984) Booming sector and Dutch disease economics: survey and consolidation. Oxford Econ Pap 36:359–380

Corden WM, Neary PJ (1982) Booming sector and de-industrialisation in a small open economy. Econ J 92:825–848. https://doi.org/10.2307/2232670

Eifert B, Gelb A, Tallroth NB (2002) The political economy of fiscal policy and economic management in oil exporting countries

Eltony MN (2002) Oil price fluctuations and their impact on the macroeconomic variables of Kuwait: a case study using VAR model for Kuwait. www.arbi-api.org/wps9908.pdf

Hamilton J (2003) What is an oil shock? J Econ 113:363–398

Hansen P.R, Lunde A (2001) A forecast comparison of volatility models: does anything beat a GARCH(1, 1)? Working Paper No. 01–04

Higgins ML, Bera AK (1992) A class of nonlinear ARCH models. Int Econ Rev 33:137–158

Johansen S (1988) Statistical analysis of cointegration vectors. J Econ Dyn Control 12:231–254

Johansen S, Juselius K (1990) Maximum likelihood estimation and inference on cointegration: with application to the demand for money. Oxford Bull Econ Stat 52(2):169–210

Lartey EKK (2006) Capital inflows, Dutch disease effects and monetary policy. Disertation for the Degree of Ph.D., Boston College

Niquidet K, Manely B (2008) Regional log market integration in New Zealand. Resource economics and policy analysis

Nwoko IC, Aye GC, Asogwa BC, Yildiz F (2016) Oil price and food price volatility dynamics: the case of Nigeria. Cogent Food Agric 2(1). https://org.doi/10.1080/23311932.2016.1142413
Pasban F (2004) Fluctuations in oil prices on agricultural production. Econ Res 4(1):117–136 (in Farsi)
Puyana A (2000) Dutch disease, macroeconomic policies and rural poverty in Colombia. Int J Politics Cult Soc 14(1):205–233
Rashidghalam M (2019) Introduction to sustainable agriculture and agribusiness in Iran. In: Rashidghalam M (eds). Sustainable agriculture and agribusiness in Iran. Perspectives on development in the middle East and North Africa (MENA) region. Springer, Singapore
Udoh E, Egwaikhide F.O (2012) Dose international oil price volatility complement domestic food price instability in Nigeria? an empirical enquiry. Int J Econ Finance 4
Von Braun J (2007) The world food situation: new driving forces and required action, food policy report. Washington, DC; IFPRI

Zahra Rasouli is a lecturer in Department of Agricultural Economics at University of Tabriz. She holds her B.Sc., M.Sc. and Ph.D. from Department of Agricultural Economics, University of Tabriz. Her field of expertise and interest are Agricultural Market Development, Time Series Modeling especially Price Seasonality and Markov Switching Modeling. She has a wide range of teaching experience in Agricultural Accounting, Agricultural Policy and Agricultural Insurance at under graduate level.

Mohammad Ghahremanzadeh is an Associate Professor in Department of Agricultural Economics at University of Tabriz. He holds a B.Sc. from University of Tabriz and M.Sc. and Ph.D. from University of Tehran. He spent six months in Australia as Research Scholar to complete his thesis at University of Queensland. His fields of expertise include Agricultural Policy, Agricultural Price Analysis, Agricultural Insurance and Risk Management and (seasonal) time series modeling. He has co-supervised over 30 M.Sc. and 4 Ph.D. students.

Masoomeh Rashidghalam is a visiting researcher at Centre for Entrepreneurship and Spatial Economics (CEnSE), Jönköping International Business School (JIBS), Jönköping, Sweden. She did her B.Sc. and Ph.D. in Department of Agricultural Economics at University of Tabriz and holds a M.Sc. from Tarbiat Modares University. Dr. Rashidghalam's areas of expertise are: Agricultural Production Economics, Productivity and efficiency, Well-Being, and Urbanization. She has a wide range of teaching experience in Econometrics, Agricultural Production Economics and Microeconomics. She has written two books: *Measurement and Analysis of Performance of Industrial Crop Production: The Case of Iran's Cotton and Sugar Beet Production*, 2018, published by Springer. *Sustainable Agriculture and Agribusiness in Iran*, 2019, published by Springer. She has publications in Journal of Productivity Analysis.

Environmental Efficiency in Agricultural Sector

Pariya Bagheri, Sahar Abedi, and Farid Bagheri Sarajug

Abstract Agriculture is an important and vital sector in Iran's economy. This sector provides basic needs along with undesirable outputs such as greenhouse gases. Greenhouse gas (GHG) emissions are main causes of environmental pollution. By considering the pollution in production process, economic performance can be evaluated precisely. We can modify the efficiency model using the desirable and undesirable outputs to improve the quality of the environment, and to achieve sustainable development. In this regard, environmental efficiency of agriculture is an appropriate guide to produce crops using low inputs and environmental pollution. This study aims to estimate the agricultural environmental efficiency in Iran's different provinces using data envelopment analysis (DEA) method. The results show that for each billion Rial of agriculture gross value added, 11.39 tons of carbon dioxide are produced. Subsequently, regarding the greenhouse gas emissions caused by proportional agricultural activity, the provinces are divided into three clusters. Mean carbon dioxide emission for each cluster is 536,663.7, 804,315.7, and 186,311.8 tons, respectively. Mean environmental efficiency for clusters is 0.294, 0.243, and 0.836, respectively. This indicates that provinces with more greenhouse gas emissions have lower environmental efficiency. In this regard, the politicians should make the less efficient clusters to reduce the use of undesirable inputs to decrease the pollutant emissions. Besides, by proper allocation of capital, the more efficient cluster can improve environmental technologies.

Keywords Environmental efficiency · Greenhouse gas · Data envelopment analysis · K-medoids clustering

---

P. Bagheri (✉) · S. Abedi
Department of Agricultural Economics, University of Tabriz, Tabriz, Iran
e-mail: pariya.ba93@gmail.com

S. Abedi
e-mail: Sahar_abedi91@yahoo.com

F. Bagheri Sarajug
Department of Accounting, Zanjan Branch, Islamic Azad University, Zanjan, Iran
e-mail: faridbagheri0@gmail.com

M. Rashidghalam (ed.), *The Economics of Agriculture and Natural Resources*,
Perspectives on Development in the Middle East and North Africa (MENA) Region,
https://doi.org/10.1007/978-981-15-5250-2_12

1 Introduction

The agricultural sector, as a provider of basic needs of the society and other economic sectors, plays a major role in production and development. To respond to the growing population demand, the gross value added of Iran's agricultural sector has risen from 300,027 (billion Rials) in 2004 to 456,018 (billion Rials) in 2018 (Central Bank of Iran 2019). Production growth is corresponding with the industrialization process. The use of greenhouses, agricultural machinery, farm products processing, selection of high-yielding varieties, applying chemical fertilizers and pesticides, and improvement of irrigation methods are examples of the agricultural industrialization process (Moutinho et al. 2018). These activities increase the energy and non-renewable fuels demands so that the fuels consumption in Iran's agricultural sector has increased from 42 million in 2008 to 51 million barrel of oil equivalent in 2017 (Statistical Center of Iran 2019). This increase in energy use damages the environment and lowers its quality (Maddison 2001).

Nowadays, regarding the importance of sustainable development, it is necessary to consider the environmental pollution caused by economic sectors. The greenhouse gas emissions caused by economic sectors are particular aspects of the environmental pollution impacts on local, regional, and global ranges (Molaei and Sani 2015). High density of these gases in the atmosphere increases Earth's global temperature. According to Abbasi et al. (2010), Faryadi (2013), Iranian Meteorological Organization (2017) and Intergovernmental Panel on Climate Change (IPCC) (2018) (Intergovernmental Panel on Climate Change (IPCC) Switzerland (2018)), within the next 50 years, the Earth's temperature will increase by about 1.5–3 °C due to 20% increase in carbon dioxide in the atmosphere. In the agricultural sector, the types of fuels, such as petrol, kerosene, gas oil, fuel oil, and natural gas, have considerable impacts on the GHG. Table 1 represents the greenhouse gas emissions of each fuel in Iran's agricultural sector in 2018. According to this table, the highest gas emission is related to carbon dioxide (11,966,481 tons), which makes the highest amount of pollution (Ministry of Energy 2018).

Besides the mentioned fuels, other agricultural inputs cause environmental pollution. In modern agriculture, there is a high correlation between the agricultural

Table 1 The greenhouse gas emissions in Iran's agricultural sector in 2018 (tons)

Fuel	N_2O	CH_4	CO_2	SPM	CO	SO_3	SO_2	NO_X
Petrol	0.02	1	564	–	83	–	–	3
Kerosene	0.40	2	42445	–	13	–	39	8
Gas oil	2951	428	7646070	18995	9498	271	44503	43418
Fuel oil	0.14	0.7	17437	5.4	0.02	4	253	54
Natural gas	8.00	76	4259965	–	–	–	–	–
Total	2959	507	11966481	19000	9594	275	44795	43483

Source Ministry of energy (2018)

products' yield and inputs, such as chemical fertilizers, pesticides, and genetically modified seeds (Pishgar-Komleh et al. 2012). Such inputs allow farmers to achieve higher yield with a lower cost per hectare, and also, increase the labor and capital efficiency (Molaei et al. 2017). According to the Food and Agricultural Organization (2019), the average use of chemical fertilizers and pesticides were 3,820,412 and 6,857.4 tons in Iran during 2010–2016, respectively.

Although some of these materials, such as nitrate fertilizers, are essential soil elements for crop growth, their excessive amounts can pollute surface water, groundwater, soil, and produce greenhouse gases that pollute the Earth's atmosphere (Kiani 2014). Consequently, in the production process, the optimal pattern of energy, chemical fertilizer, and pesticide consumption needs to be used to minimize environmental pollution. In other words, besides the economic efficiency the environmental effects must be considered to achieve sustainable growth and development.

In the environmental pollution impacts' studies, a group of studies including Ball et al. (2004), Darijani et al. (2005), Murty and Kumar (2006), Nanere et al. (2007), Nguyen et al. (2008), Rezaei et al. (2012), Jafarian and Esmaeli (2013), Kiani (2014), Molaei and Sani (2015), Molaei et al. (2017) and Sintori et al. (2019) concluded that the environmental conditions determine the firms' efficiency significantly. Hence, the environmental efficiency must be considered to obtain a realistic view of firms' situation.

Another group of the studies on environmental efficiency considered the Kuznets' curve. Pajooyan and Moradhasel (2008), Pourkazami and Abrahimi (2008), Bargi Oskuyi and Yavâri (2008), Vaseghi and Esmaeili (2010), Dargahi and Bahrami Gholami (2012), Harati et al. (2016), Salmani et al. (2017) and Fatahi et al. (2018) indicated that there is a significant relationship between the carbon dioxide emission and per capita income in Iran. Most of the studies in the environmental efficiency field focused on the energy sector and the pollution caused by fossil fuels. (e.g., see Wang et al. 2012; Asghari and Salarnazar Rafsanjanipour 2013; Goto et al. 2014; Sajadifar et al. 2016; Parsa et al. 2016). Khoshnevisan et al. (2013), Alhami et al. (2017) and Arefi et al. (2018) investigated a special region in Iran and quantified the effect of undesirable inputs, such as chemical fertilizers, pesticides and fossil fuels on greenhouse gas emissions. The reduction of greenhouse gas emissions were calculated using undesirable inputs reduction. However, they did not consider the greenhouse gas emissions as undesirable outputs. Hence, in this study, we try to consider the effect of all undesirable inputs, i.e., chemical fertilizers (nitrogen, phosphate, potassium), pesticides (fungicides, pesticides, herbicides), and fossil fuels (natural gas, petrol, gas oil, fuel oil, kerosene), besides, and undesirable outputs.

According to the discussed issues, agricultural inputs which are used to get more production and different kinds of energy, cause lots of pollution, and cause different greenhouse gas emissions. Hence, to conduct a precise evaluation of this sector's performance and efficiency, it is essential to consider the environmental aspects of agricultural activities. Without such parts we cannot reach the sustainable development in the agricultural sector. In this study, to estimate the agricultural economic-environmental efficiency of different provinces, we consider the pollution that arise from energy use and chemical fertilizers and pesticides usage. To achieve the aim of study, we use data envelopment analysis (DEA) method. Furthermore, regarding the greenhouse gas emissions caused by agricultural activities, we clustered the provinces of Iran using the k-medoids algorithm. Such clustering and efficiency evaluations will provide an accurate view to adopt appropriate programs and policies so that the sustainable development will be available.

The study is organized as follows: Sect. 2 presents the proposed methodology, followed by Sect. 3 that reports the results of the study. Finally, Sect. 4 summarizes and concludes, including some policy recommendations to policymakers.

2 Methodology

In this study, considering the greenhouse gas emissions caused by agricultural activities, we clustered provinces using k-medoids algorithm to calculate the agricultural environmental efficiency in Iran. Clustering is an efficient method in scientific and managerial research. It divides a dataset in d-dimensional space so that the similarity within a cluster is maximal and the compatibility between clusters is minimal. Consider n-object population which is described by m features and divided into k clusters. $X_i = (x_{i1}, x_{i2}, \ldots, x_{im})$ denotes the m features vector of the object i, and $X = \{x_1, x_2, \ldots, x_n\}, x_i \in R^m$ represents the database. There are different types of clustering methods that are used in a wide range of contexts. Among the clustering algorithms, the k-means is a commonly used method to classify the community that seeks a nearly optimal clustering with a specified clusters' number. K-means is a non-hierarchical clustering method that uses the iterative hill-climbing algorithm (Kim and Ahn 2008). The steps of the k-means clustering algorithm implementation are as follows:

1. In each cluster, a point is randomly selected as the centroids of initial clusters.
2. According to the distance criterion, the data are allocated to the centroids. The nearness criterion is Euclidean distance:

$$d(x_i, c_j) = \sqrt{(\sum_{i=1}^{c} (x_{ik} - c_{jk})^2} \tag{1}$$

where x_i is the object i and c_j is the centroid of the cluster j. The input data are allocated to a cluster that has the minimum distance with the cluster centroid. The centroids are equal to the arithmetic mean of the cluster's data.

3. Update the clusters' centroids, i.e., compute the average of each cluster.
4. Repeat steps 2 and 3 until the clusters become fixed (Mohammadi and Rouhani 2017).

The k-means method is applied in different branches of science due to its simplicity and rapid convergence. However, this method is suitable when the number of members in each cluster is almost the same. However, this method is sensitive to the outliers so that it does not cope with overlapping clusters, and the clusters can be exited from the centroids by outliers (Kim and Ahn 2008). To overcome this problem, we use the k-medoids clustering algorithm. In this method, medoids are used instead of cluster centroid. The k-medoids uses the medoid instead of the average of each cluster. In a cluster, medoid is the most centrally located object. Since this method is based on the indicator points or medoids, it has less sensitivity to outliers than k-means method (Park and Jun 2009). Primary strategy of k-medoid method is finding k clusters for n objects by detecting an arbitrary indicator point for each cluster. Rest of the objects that are similar to the clusters' medoids are divided into clusters. The k-medoids general algorithm is as follows (Velmurugan and Santhanam 2011):

Input	k: number of clusters
	D: The dataset of n objects
Output	k clusters that has the minimum sum of the dissimilation of all objects associated with their closest medoids
Method	Arbitrarily select k objects in D as the initial indicator points
Repeat	Allocate each object to the cluster that has the nearest medoid Randomly choose a non-medoid object O_{random} Calculate the total objects O of swapping object O_j with O_{random} If $O < 0$, swap O_j with O_{random} to create a new set of k modid Repeat this steps until changes stop

Among the various k-medoids clustering algorithms, partitioning around medoids (PAM) proposed by Kaufman and Rousseeuw (1990) is the most efficient one (Park and Jun 2009). This algorithm seeks a sequence of objects called medoids, and is a form of steepest ascent hill-climbing with a simple swap neighborhood operation. In the PAM method, k is selected in advance. The objects that are called medoids are placed in a set of selected objects S. O is a set of objects, and $U = O\text{-}S$ is a set of unselected objects. In each repetition of the loop, we select medoid objects and non-medoid objects, and the best clustering is formed when their roles are changed. The objective function corresponds to the sum of distances from each object to the closet medoid, and must be minimized. The purpose of this method is to minimize the objects' dissimilarity average to their nearest medoid. Similarly, we can also minimize the sum of the objects' dissimilarities. This algorithm has two steps:

1. First step, BUILD: a set of k objects are chosen for the initial set of selected objects.
2. Second step, SWAP: an attempt to improve the clustering quality by swapping selected points with unselected points (UMASS Boston 2018; Reynolds et al. 2004).

Kaufman and Rousseeuw (1990) introduced the silhouette coefficient to determine the objects of each cluster and the number of clusters. For each element or object i, suppose $a(i)$ is the mean distance of i from rest objects in cluster C_i, and for other cluster $C \neq C_i$, suppose $d(i, C)$ is the mean distance of object i from the elements in C. After calculating $d(i, C)$ for all clusters, assume that $b(i)$ is the smallest value. The cluster, which is obtained this minimum value, is called the neighbor of object i. The silhouette coefficient is determined as follows:

$$s(i) = \frac{b(i) - a(i)}{\max\{a(i), b(i)\}} \tag{2}$$

If $s(i)$ is close to 1, i is properly classified, and if it is close to 0, it is not clear whether i belongs to C_i or to its neighbor. Negative values also indicate a misclassification. The mean $s(i)$ for all elements is called the mean silhouette coefficient for the entire data that its maximum value is used to select an appropriate number of clusters (Reynolds et al. 2004).

Efficiency is defined as optimal allocation of resources, and indicates the ratio of the observed output to the maximum output specified by the production frontier. It is compression between the performances of a firm and the best firm of an industry (De Koeijer et al. 2002). In general, two techniques are used to measure the efficiency, i.e., stochastic frontier analysis (parametric approach) and data envelopment analysis (non-parametric approach) (Jarzebowski et al. 2013), Rashidghalam and Heshmati 2019). Data envelopment analysis (DEA) is a non-parametric method based on linear programming to determine the efficiency of homogeneous economic firms (De Koeijer et al. 2002).

DEA is divided into input and output-oriented approaches under the constant and variable returns to scale assumptions. The input-oriented approach minimizes the inputs on condition that output is constant. In contrast, output-oriented approach maximizes the output on condition that inputs are constant (Sajadifar et al. 2016). Standard DEA model assumes that decrease in inputs and increase in output can improve the efficiency, while in the real production process, undesirable outputs, such as carbon dioxide which pollutes the environment, may be are released during the input conversion into output. Therefore, the undesirable output must be decreased to improve the efficiency (Chiu et al. 2016; Sajadifar et al. 2016). Hence, the environmental efficiency is an aspect of technical efficiency that focuses on the inputs with adverse environmental effects (energy, chemical fertilizers, and pesticides). Thus, environmental efficiency is defined as the production of goods and services with minimum energy and input use that cause less pollution as well. The general DEA model is:

$$\begin{aligned} &\min_{\theta,\lambda} \ \theta \\ &s.t. \ \ \theta - X\lambda \geq 0 \\ &\quad\ \ y_i - Y\lambda \leq 0 \\ &\lambda : \begin{cases} \lambda \geq 0 \ CRS \\ e\lambda = 1 \ VRS \end{cases} \end{aligned} \tag{3}$$

where X and Y are the input and output matrices, respectively. θ denotes the efficiency of province i. If $\theta = 1$, then i is a point on the production frontier that indicates the province i is efficient. While, if $\theta \leq 1$, and then by improving the inputs, the efficiency increases. λ is a vector of constant coefficients where $\lambda \geq 0$ and $e\lambda = 1$ represent the constant and variable returns to scale, respectively (Graham 2004; Paradi et al. 2015). In this study, due to the scarcity of resources and different economic sizes of provinces, we used the input-oriented approach and variable returns to scale. According to Schmidheiny and Zorraquin (1996), Ishikawa and Huppes (2005), and Moutinho et al. (2018), the agricultural economic efficiency is the ratio of agricultural gross value added to the greenhouse gas emissions. According to Table 1, since carbon dioxide causes the maximum pollution in Iran, we select this gas as an undesirable output. Consequently, to consider the desirable and undesirable outputs, their ratio is entered to the DEA model as:

$$Y = \frac{Y_{GVA}}{Y_{CO_2}} \tag{4}$$

where Y_{GVA} and Y_{CO_2} denote the agricultural gross value added at constant prices (billion Rials) and the carbon dioxide emissions (tons), respectively, caused by agricultural activities. The amount of carbon dioxide emissions is not available for each province in agricultural sector; therefore, the amount of carbon dioxide emission is calculated as follows:

$$CE_{ff} = \sum_{i=1}^{n} FC_i * EF_i \tag{5}$$

where CE_{ff} is the amount of carbon dioxide emissions, FC_i is the amount of fuel and chemical fertilizers and pesticides, and EF_i is carbon dioxide emission coefficient that is presented in Table 2 (Ke et al. 2013; Parsa et al. 2016).

Table 2 Emission coefficient of carbon dioxide gas

Input	CO_2 emission coefficient	Reference
Nitrogen fertilizer (kg)	1.3000	(Lai, 2004), Alhami et al. (2017), Arefi et al. (2018)
Phosphate fertilizer (kg)	0.2000	(Lai, 2004), Alhami et al. (2017), Arefi et al. (2018)
Potassium fertilizer (kg)	0.2000	(Lai 2004), Alhami et al. (2017), Arefi et al. (2018)
Fungicides (kg)	3.9000	(Lai 2004), Arefi et al. (2018)
Pesticides (kg)	5.1000	(Lai 2004); Arefi et al. (2018)
Herbicides (kg)	6.3000	(Lai 2004), Arefi et al. (2018)
Natural gas (kg/m^3)	1.8979	Khodadadkashi et al. (2016), IPCC (2018)
Petrol (liter/kg)	2.2898	Khodadadkashi et al. (2016), IPCC (2018)
Kerosene (liter/kg)	2.5566	Khodadadkashi et al (2016), IPCC (2018)
Gas oil (liter/kg)	2.6847	Khodadadkashi et al. (2016), IPCC (2018)
Fuel oil (liter/kg)	3.0013	Khodadadkashi et al. (2016), IPCC (2018)

3 Results and Discussion

The input vector includes labor force (number of employees), the agricultural capital[1] (billion Rials), land (hectare), chemical fertilizers, including, nitrogen, phosphate, and potassium (ton), chemical pesticides, including, fungicides, pesticides, and herbicides (liter/kg), and energy, including, natural gas, petrol, gas oil, fuel oil, and kerosene (million liters) (Ministry of Energy 2018, Ministry of Agriculture-Jihad 2018).

To evaluate the agricultural environmental efficiency for different provinces, the data were collected in 2018 and their descriptive results are shown in Table 3. Investigation of data shows that the agricultural sector's gross value added is about 1,178,641 billion Rials, and this sector produces 13,420,259.4 tons of carbon dioxide (Statistical Center of Iran 2019); in other words, for each billion Rial of agriculture gross value added, 11.39 tons of carbon dioxide is released. The maximum amount of agricultural value added, and agricultural capital is related to Fars Province and the minimum amount to Qom Province. The highest and lowest amounts of carbon dioxide emissions are associated with Fars and Bushehr Provinces, respectively.

[1]Capital stock in agricultural sector is calculated using the capital–output ratio method. In this method, it is assumed that the ratio of agricultural capital to total capital for each province is equal to the ratio of agricultural gross value added to total gross value added. Hence, the provinces' agricultural capital is equal to the product of the rate and total capital.

Table 3 The data description in Iran's provinces

Inputs		Min	Max	Average	Total
Labor force (number of employees)		Qom	Razavi Khorasan	166332	5156306
		14632	609271		
Agricultural capital (billion Rails)		Qom	Fars	1428	44268
		385.2	4069.4		
Land (hectare)		Alborz	Razavi Khorasan	531503.5	16476609
		50227	1460111		
Chemical fertilizers (ton)	Nitrogen	Alborz	Khuzestan	42367.65	1313397
		2488	191867		
	Phosphate	Alborz	Khuzestan	10963.03	339854
		885	36672		
	Potassium	Alborz	Golestan	3162.3	98032
		223	13372		
	Total	Alborz	Khuzestan	56493	1751283
		3596	230946		
Chemical pesticides (liter/kg)	Fungicides	–	East Azerbaijan	7702.7	53919
		–	21309		
	Pesticides	–	East Azerbaijan	6072.8	36437
		–	10744		
	Herbicides	–	Zanjan	4097.2	20486
		–	5890		
	Total	–	Mazandaran	23125.3	346880
		–	116024		
Energy	Natural gas (million m^3)	Sistan and Baluchestan	Mazandaran	63.54	1969.5
		0	264.8		
	Petrol	–	Khuzestan	118.5	237
	(1000 L)	–	199		
	Gas oil	Kohgiluyeh and Boyer-Ahmad	Fars	87535.9	2713614
	(1000 L)	4061	292759		
	fuel oil	–	Fars	902.2	5413

(continued)

Table 3 (continued)

Inputs		Min	Max	Average	Total
	(1000 L)	–	2550		
	Kerosene (1000 L)	–	Mazandaran	707.1	16263
		–	14125		
	Total	Sistan and Baluchestan	Mazandaran	63620.5	1972236
	(Million liters)	147.34	264912.2		
Agriculture gross value added (billion Rails)		Qom	Fars	38020.7	1178641
		9342	102046		
Carbon dioxide emission (tone)		Bushehr	Fars	432911.5	13420259
		43890	1082453		

The maximum amount of pesticides, fertilizers, and energy consumption are related to Khuzestan and Mazandaran Provinces, respectively. As shown in Table 3, the amounts of allocated inputs to the agricultural sector, including land, labor, capital, energy, pesticides, and fertilizer are 16,476,609, 5,156,306, 44,268.06, 1,972,236, 346,880, and 1,751,283, respectively.

Nest, regarding the greenhouse gas emissions caused by agricultural activities, the provinces are clustered using k-medoids algorithm. For this purpose, the distance matrix is measured using the Euclidean distance method, and the results are presented in Fig. 1. In Fig. 1, numbers from 1 to 31[2] refer to different provinces, and the distance based on colors is determined in the right. The elements on the distance matrix diameter show the distance of each province data with itself that it is zero. The provinces that are specified by dark blue color have considerable difference amounts of carbon dioxide emissions, and the provinces with orange color have less difference.

According to the distance matrix, there are three parts: 1. Blue parts are maximal distances, 2. orange parts denote minimal distances, and 3. White parts are moderate distances. Therefore, it seems that three clusters are suitable to divide the provinces. However, since the optimal number of clusters is essential in k-medoids clustering, the mean silhouette coefficient is used to assure. To determine the optimal clusters' number, we measured the values of criteria for different numbers of clusters and compared them. The mean silhouette coefficient is presented in Table 4. The higher value of criteria is important, therefore, the optimal number of clusters is 3. Figure 2 shows the silhouette graph, which indicates that number 3 is the peak point of the graph.

---

[2] 1. East Azerbaijan, 2. West Azerbaijan 3. Ardabil, 4. Isfahan, 5. Alborz, 6. Ilam, 7. Bushehr, 8. Tehran, 9. Chaharmahal and Bakhtiari, 10. South Khorasan, 11. Razavi Khorasan, 12. North Khorasan, 13. Khuzestan, 14. Zanjan, 15. Semnan, 16. Sistan and Baluchestan, 17. Fars, 18. Qazvin, 19. Qom, 20. Kurdistan, 21. Kerman, 22. Kermanshah, 23. Kohgiluyeh and Boyer-Ahmad, 24. Golestan, 25. Gilan, 26. Lorestan. 27. Mazandaran, 28. Markazi, 29., 30. Hamedan, 31. Yazd.

Fig. 1 Distance matrix of provinces in terms of carbon dioxide emissions caused by agricultural sector

Thus, considering the optimal number of clusters, the k-medoids method is estimated, and its results are reported in Table 5. Table 5 shows the medoids or indicator points of each cluster, the clustering vector of the provinces based on the nearness criterion to the cluster medoid, and the minimum values of objective function for BUILD and SWAP steps. Therefore, according to the carbon dioxide emissions caused by agricultural sector, the provinces are divided into three groups that their distribution is shown in Fig. 3. In Fig. 3, using the silhouette coefficient, each province is allocated to the nearest medoid. The silhouette coefficient of each province is positive and significantly different from zero. This indicates that the provinces are properly clustered.

According to Fig. 3, first and third clusters include 13 provinces, and second cluster contains 5 provinces. The averages of carbon dioxide emissions from agricultural activities in clusters are 536,663.7, 804,315.7, and 186,311.8 tons, respectively. Regarding the clusters' medoids in Table 5, it is also clear that the provinces of the second cluster have used the maximum amount of inputs and have generated the highest amount of gross value added.

After clustering the provinces and identifying each cluster members, the agricultural environmental efficiency of provinces are measured using the DEA with input-oriented and variable returns to scale approach. The results of environmental

Table 4 The mean silhouette coefficient for different number of clusters

Number of clusters	2	3	4	5	6	7	8	9	10
Mean of silhouette	0.439	0.487	0.445	0.453	0.418	0.445	0.453	0.439	0.369

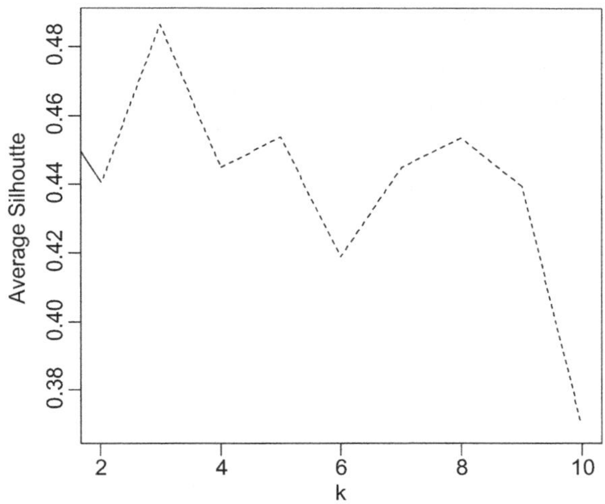

Fig. 2 The silhouette graph to determine the optimal number of clusters

Table 5 k-medoids results for clustering provinces based on carbon dioxide emissions caused by agricultural sector

Medoids

ID	Gross value added	CO_2	Land	Labour	Capital	Energy	Fertilizer
30	40243	485932.8	872174	165497	1657.661	34635.2	58374
4	50523	877032.7	382467	213041	2024.985	201055.3	57270
9	19749	96739.85	182112	72463	681.182	16114.8	25159

Clustering vector

Province (1, 2, 3, 4, 5, 6, 7, 8, 9, 10, 11, 12, 13, 14, 15, 16, 17, 18, 19, 20, 21, 22, 23, 24, 25, 26, 27, 28, 29, 30, 31)

k(1, 1, 1, 2, 3, 3, 3, 2, 3, 3, 1, 3, 1, 1, 3, 3, 1, 3, 3, 1, 1, 1, 3, 2, 3, 1, 2, 1, 3, 1, 2)

Objective function

BUILD	271304.1	SWAP	249997

efficiency are shown in Table 6. As shown in this table, the averages of environmental efficiency in the first, second, and third clusters are 0.294, 0.243, and 0.836, respectively. Hence, it is found that the second and third clusters with the highest and lowest carbon dioxide emissions, respectively, have the minimum and the maximum environmental efficiency.

The results of the provinces clustering in Table 6 show that in the third cluster, the environmental efficiency of agriculture in the provinces of Alborz, Bushehr, South Khorasan, Sistan and Baluchestan, Kohgiluyeh and Boyer-Ahmad, and Hormozgan are equal to 1. Compared to other provinces, it can be claimed that these provinces have fewer facilities, and lower capital (on average, 746.27 billion Rials) is allocated to them. However, due to providence approach, optimized use of inputs and efficient

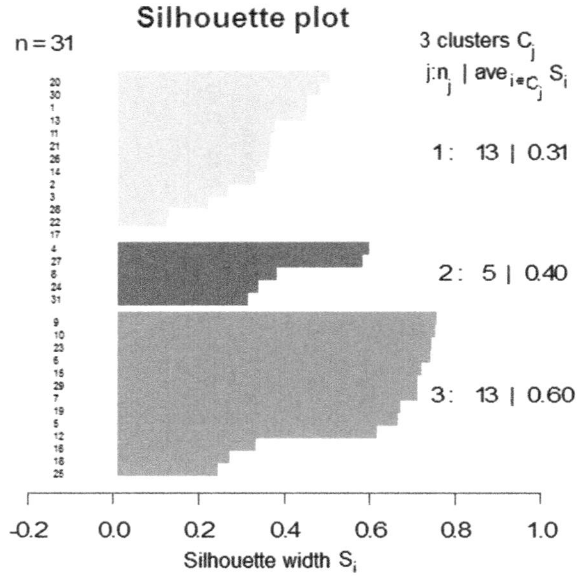

Fig. 3 Provinces clustering based on the carbon dioxide emission using the k-medoids and the silhouette coefficient

Table 6 Results of clusters' agricultural environmental efficiency

Province	Environmental efficiency	Province	Environmental efficiency	Province	Environmental efficiency
East Azerbaijan	0.209	Isfahan	0.194	Alborz	1.000
West Azerbaijan	0.299	Tehran	0.287	Ilam	0.882
Ardabil	0.319	Golestan	0.315	Bushehr	1.000
Razavi Khorasan	0.141	Mazandaran	0.159	Chaharmahal and Bakhtiari	0.804
Khuzestan	0.218	Yazd	0.262	South Khorasan	1.000
Zanjan	0.432			North Khorasan	0.664
Fars	0.170			Semnan	0.875
Kurdistan	0.423			Sistan and Baluchestan	1.000
Kerman	0.232			Qazvin	0.340
Kermanshah	0.391			Qom	1.000
Lorestan	0.376			Kohgiluyeh and Boyer-Ahmad	1.000
Markazi	0.34			Gilan	0.306
Hamedan	0.281			Hormozgan	1.000
Average	0.294	Average	0.243	Average	0.836

management, these provinces release less pollution, and have more environmental efficiency. Environmental regulations will not affect the production technology of these provinces. In contrast, the provinces that are located in the second cluster are in a better position in terms of facilities and allocated capital (on average, 1,896.81 billion Rials). Nevertheless, they do not use the inputs efficiently, and produce more carbon dioxide than other provinces. In fact, poor provinces compared to rich ones, generate more value added than releasing carbon dioxide. Investigation of inputs shows that the second cluster uses more undesirable inputs than two other clusters, which could increase the carbon dioxide emissions and degrade environmental efficiency. Thus, given the present technologies the provinces, that their agricultural environmental efficiency are lower than 1, and generally are located in clusters 1 and 2, can significantly reduce the inputs usage to decrease the carbon dioxide emissions and to generate the same gross value added.

The results of agricultural environmental efficiency show that Razavi Khorasan, Mazandan, and Fars Provinces have the least efficiency scores. Razavi Khorasan Province is the greatest producer of saffron in the country that produces more than 80% of the country's saffron (Ministry of Agriculture Jihad 2018). According to Table 3, this province has the maximum land and labor force. In term of land scale, studies by Golkaran Moghadam (2013) and Mahmoodabadi (2012) showed that smallholder was an important reason of inefficiency in agricultural sector. Mazandan Province is the country's rice production center and according to studies by Eskandari et al. (2011), Mansoori et al. (2012), Pour Sherazi et al. (2013) and Table 3, uses high levels of energy and chemical pesticides in rice production. According to Table 2, Fars Province has the highest value added and greenhouse gas emissions. This province also uses the maximum amount of gas oil so that its efficiency score is low. In addition, this province is one of the biggest producers of red and white meat (Ministry of Agriculture jihad (2018) reported that this province is ranked second in red meat and sixth in white meat production) and according to the studies by Tavakoli et al. (2014) their waste repels is traditionally done and increases the greenhouse gas emissions.

4 Conclusion and Policy Recommendations

In this study, the provinces of Iran are divided into three clusters using k-medoids method in terms of carbon dioxide emissions caused by agricultural activities. The amounts of carbon dioxide emissions in three clusters are 536,663.7, 80,4315.7, and 186,311.8 tons, respectively. The provinces of second cluster, which release the highest carbon dioxide amount, have a higher undesirable inputs use than two other clusters. The results of environmental efficiency using the data envelopment analysis show that the averages of environmental efficiency of agriculture in these clusters are 0.294, 0.243, and 0.836, respectively. It is clear that the clusters with more use of undesirable inputs also release more carbon dioxide, and consequently, have less environmental efficiency.

According to the clustering results, the first cluster's provinces use the maximum amount of land however, their environmental efficiency is lower than the third cluster; this indicates that land is not used efficiently in these provinces. In addition, most of these provinces suffer from smallholder. Consequently, consolidation of land, use of small-scale technologies, and education of farmers about optimal input use can improve the agricultural environmental efficiency in the first cluster. Furthermore, waste management methods in poultry farms and slaughterhouses are appropriate mechanism to reduce wastes. In the second cluster, the provinces have the highest use of chemical fertilizers, pesticides, and energy. These undesirable inputs generate more carbon dioxide as an undesirable output that decreases the environmental efficiency. In this cluster, semi-mechanized cultivation method can reduce energy consumption and increase the efficiency. Besides, integrated pest management methods and organic fertilizers are useful tools in greenhouse gas emissions reduction.

By clustering the provinces in terms of greenhouse gas emissions and calculating their agricultural environmental efficiency, the policymakers can provide a separate plan for each cluster. Hence, by defining the environmental regulations, the politicians can obligate the less efficient provinces to reduce the use of energy and fertilizer inputs, encourage them to apply healthier technologies, and decrease the pollutant emissions. Besides, by supporting packages and proper allocation of capital, the more efficient provinces can improve environmental technologies.

References

Abassi F, Habibi Nokhandan M, Goli Mokhtari L, Malbousi Sh (2010) Climate change assessment over Iran in the future decades using MAGICC-SCENGEN model. Phy Geogr Res Q 42(72):91–109

Alhami B, Akram A, Khanali M (2017) Optimization of energy consumption and mitigation of greenhouse gas emission of irrigated lentil production using data envelopment analysis. Iran J Syst Eng 47(4):701–710

Arefi R, Soltani A, Ajam Norozei H (2018) Carbon dioxide emission and global warming potential of energy consumption in the cotton production in golestan province. J Agroecology 10(2):529–546

Asghari M, Salarnazar Rafsanjanipour S (2013) The impact of foreign direct investment inflow in selected MENA countries. Environ Qual Econ Dev Res 3(9):1–30

Ball VE, Lovell CAK, Luu H, Nehring R (2004) Incorporating environmental impacts in the measurement of agricultural productivity growth. J Agric Resour Econ 29(3):436–460

Bargi Oskuyi M, Yavari K (2008) The impact of trade liberalization on the greenhouse gases (co2 emission) in EKC. J Econ Res 43(1):1–21

Central Bank of Iran (2019) National accounts. https://www.cbi.ir

Chiu Y, Shyu M, Lee J, Lu Ch (2016) Undesirable output in efficiency and productivity: example of the G20 countries. Energy Sources Part B 11(3):237–243

Dargahi H, Bahrami Gholami M (2012) The GHGs emission determinates in selected OECD and OPEC countries and the policy implications for Iran: (panel data approach). J Iran Energy Econ 1(1):73–99

Darijani A, Sharzei Gh, Yazdani S, Paykani Gh, Sadrashrafi M (2005) Estimation of environmental efficiency by stochastic frontier analysis a case study of livestock slaughterhouses in Tehran. Agric Econ Dev 13(51):113–134

Eskandari F, Bahrami H, Asakereh A (2011) Energy servey of mechaized and traditional rice production system in Mazandaran province of Iran. Afr J Agric Res 11:2565–2570

Faryadi S (2013) Environmental management Municipality Organization. https://www.noandishaan.com

Fatahi Sh, Heidarian M, Moradi S (2018) The technical and scale effects of economic growth on the environmental in Iranian provinces; panel var approach. Q Energy Econ Rev 14(57):147–171

Food and Agricultural Organization (2019) Data, inputs, fertilizers, and pesticides archive. https://www.fao.org

Golkaran Moghadam S (2013) Comparison and analysis efficiency of saffron farmers in selected township of Khorasan Razavi province. Agric Econ Dev 21(84):79–101

Goto M, Otsuka A, Sueyoshi T (2014) DEA (Data Envelopment Analysis) assessment of operational and environmental efficiencies on Japanese regional industries. Energy 66:535–549

Graham M (2004) Environmental efficiency: meaning and measurement and application to Australian dairy farms. In: Presented at the 48th annual AARES conference melbourne, Victoria

Harati J, Taghizadeh H, Amini T (2016) Investigating the impact of trade and political variable on environmental performance index: a dynamic panel analysis. J Econ Policy 7(14):130–157

Huppes G, Ishikawa M (2005) A Framework for Quantified Eco-efficiency Analysis. J Ind Ecol 9(4):25–41

I.R. of Iran Meteorological Organization, Atmospheric Science & Meteorological Research Center (ASMERC), Climatological Research Institute (2017) revelation-assessing the climate change impacts and perspectives of the 21st century in Iran, 1–45

Intergovernmental Panel on Climate Change (IPCC) (2018) Sixth assessment report—climate change. https://www.ipcc.ch

Intergovernmental Panel on Climate Change (IPCC) Switzerland (2018) Global warming of 1.5 °C. https://www.ipcc.ch

Jafarian M, Esmaeli A (2013) Application of environmental impact analysis of technical efficiency: a case study of Shiraz fattening units of beef cattle. J Agric Econ Res 5(2):151–164

Jarzebowski S, Jarzebowski A, Klepacki B (2013) Efficiency and integration in the food supply Chain. J Food Syst Dyna 4(3):159–169

Kaufman L, Rousseeuw P (1990) Finding groups in data: an introduction to cluster analysis. Wiley Series in probability and statistics

Ke J, McNeil M, Price L, Zheng Khanna N, Zhou N (2013) Analysis and practices of energy benchmarking for industry from the prospective of systems engineering. Energy 54(c):32–44

Khodadadkashi F, Akabari Tafti M, Mosavi Y, Khosravineghad A (2016) Calculating the social costs of carbon dioxide emissions in different provinces of Iran. Q J Energy Policy Plann Res 2(2):77–110

Khoshnevisan B, Rafiee Sh, Omid M, Mousazadeh H (2013) Reduction of CO2 emission by improving energy use efficiency of greenhouse cucumber production using DEA approach. Energy 55(2013):676–682

Kiani GH (2014) Environmentally adjusted productivity measurement of the Iranian agricultural sector. Environ Sci 11(4):63–72

Kim K, Ahn H (2008) A recommender system using GAK-means clustering in an online shopping market. Expert Syst Appl 34:1200–1209

De Koeijer TJ, Wossink GA, Struik PC (2002) Measuring agricultural sustainability in terms of efficiency: the case of Dutch sugar beet growers. J Environ Manag 66:9–17

Lai R (2004) Carbon emission from farm operations. Environ Int 30(7):981–990

Maddison A (2001) The word economy, a millennial perspective. Development Centre Studies OECD Paris

Mahmoodabadi Z (2012) Razavi Khorasan and its problems. The Center for Strategic Studies, 1–126

Mansoori H, Rezvani Moghadam P, Moradi R (2012) Energy budget and economic analysis in conventional and organic rice production systems and organic scenarios in the transition period in Iran. Energy 6:341–350

Ministry of Agriculture-Jihad Statistic information, agricultural statistical yearbook (2018). https://www.maj.ir
Ministry of Energy (2018) Energy balance sheet deputy for planning and economic affairs. https://www.pep.moe.gov.ir
Mohammadi B, Rouhani AK (2017) Application of k-means, fuzzy c-means and Gustafson-Kessel FCM methods in integration of refraction seismic tomography and electrical resistivity data inversion results for evaluation of the alluvium and bedrock. Kharazmi J Earth Sci 3(2):183–198
Molaei M, Hesari N, Javanbakht O (2017) The estimation of input-oriented environmental efficiency of agricultural products (case study: environmental efficiency of rice production). J Agric Econ 11(2):157–172
Molaei M, Sani F (2015) Estimating environmental efficiency of the agricultural sector. J Agric Sci Sustain Prod 25(2):92–101
Moutinho V, Robaina M, Macedo P (2018) Economic-environmental efficiency of European agriculture–a generalized maximum entropy approach. Agric Econ 64(10):423–435
Murty MN, Kumar S, Paul M (2006) Environmental regulation, productive efficiency and cost of pollution abatement: a case study of the sugar industry in India. J Environ Manag 79:1–9
Nanere M, Frase I, Quazi A, Souza C (2007) Environmentally adjusted productivity measurement: an Australian case study. J Environ Manag 85:350–362
Nguyen VH, Shashi K, Virginia M (2008) Shadow prices of environmental outputs and production efficiency of household- level paper recycling units in Vietnam. Ecol Econ 65:98–110
Pajooyan J, Moradhasel N (2008) Assessing the relation between economic growth and air pollution. Tarbiat Modares Univ Press Econ Res 7(4):141–160
Paradi JC, Min E, Yang X (2015) Evaluating Canadian bank branch operational efficiency from staff allocation: A DEA approach. Manag Organ Stud 2(1):52–65
Park HS, Jun CH (2009) A simple and fast algorithm for K-medoids clustering. Expert Syst Appl 36:3336–3341
Parsa P, Jalaei A, Sadegi Z (2016) Analysis of environmental technical efficiency in the provinces of Iran. J Environ Nat Resour Econ 1(2):81–103
Pishgar-Komleh SH, Keyhani A, Mostofi-Sarkari MR, Jafari A (2012) Energy and economic analysis of different seed corn harvesting systems in Iran. Energy 43:469–476
Pour Sherazi S, Rasam G, Rajabi M (2013) Investigation of environmental impacts and energy consumption in rice production system. In: The 1st conference and exhibition on environmental and clean industry, 1–7
Pourkazami M, Abrahimi E (2008) Examining environmental kuzentes curve in Middle East. Iran J Econ Res 10(34):57–71
Rashidghalam M, Heshmati M (2019) A comparison of different full and partial nonparametric frontier models for measuring technical efficiency: with an application to Iran's cotton producing provinces. J Agric Crop Res 7(6):82–94
Reynolds AP, Richards G, Rayward-Smith V.J (2004) The application of k-medoids and PAM to the clustering of rules. In: 5th international conference on intelligent data engineering. Lecture notes in computer science 3177:173–178
Rezaei A, Amade H, Mohamadi T (2012) Policy analysis of natural gas export to India-Pakistan countries, a game theoretic approach. J Iran Energy Eco 1(2):93–126
Sajadifar S, Asali M, Fathi B, Mohamadbagheri A (2016) Measuring energy consumption efficiency using data envelopment analysis (DEA) with undesirable factors. J Plann Budgeting 20(4):55–70
Salmani M, Shokri M, Abedzadeh K (2017) Study factors affecting emission of gas co2 in Iran. Iran J Energy 20(1):55–74
Schmidheiny S, Zorraquín F (1996) Financing change the financial community, eco-efficiency, and sustainable development. Jonathan LashPresident, World Resources Institute
Sintori A, Liontakis A, Irene Tzouramani I (2019) Assessing the environmental efficiency of Greek dairy sheep farms: GHG emissions and mitigation potential. Agriculture 9(28):2–14
Statistical Center of Iran Presidency of the I.R.I Plan and Budget Organization (2019). https://www.amar.org.ir

Tavakoli M, Mosavi N, Tahari F (2014) Economic analysis of broiler farms in fars province with emphasis on environmental consideration 6(24):39–54

UMASS Boston (2018) Department of computer science. College of Science and Mathematics. https://www.cs.umb.edu/cs738/pam1.pdf

Vaseghi E, Esmaeili A (2010) Investigation of the determinant of co2 emission in Iran (using environmental Kuznets curve). J Environ Stud 35(52):99–110

Velmurugan T, Santhanam T (2011) A survey of partition based clustering algorithms in data mining: an experimental approach. Inf Technol J 10(3):478–484

Wang Q, Zhou P, Zhou D (2012) Efficiency measurement with carbon dioxide emissions: the case of China. Appl Energy 90:161–166

Pariya Bagheri holds B.Sc. in Agricultural Economics and M.Sc. in Production Economics in Department of Agricultural Economics, University of Tabriz. Her main research interest centers on Production Economics, Risk Management and Climate Change.

Sahar Abedi holds B.Sc. and M.Sc. in Agricultural Economics from Department of Agricultural Economics, University of Tabriz. For her MA Thesis, she aimed to have a better understanding of weather-based crop insurance premium for wheat crop. Her research interest lies at Risk Management and Crop insurance.

Farid Bagheri Sarajug holds B.Sc. in Accounting from Islamic Azad University, Tabriz Branch. Currently, he is a student for master degree in Accounting in Department of Accounting, Islamic Azad University, Zanjan Branch. His main research interest centers on Sustainable Development. He is financial manager and offers tax consultancy services to trading companies.

Author Index

A
Abedi, Sahar, 91, 183

B
Bagheri, Pariya, 91, 183

D
Dourandish, Arash, 49

F
Fantke, Peter, 11, 21

G
Ghahremanzadeh, Mohammad, 171

H
Haghjou, Maryam, 109, 125
Hassani, Leila, 11, 21
Hayati, Babollah, 31, 109, 125

K
Kakhki, Mahmoud Daneshvar, 11, 21, 49
Kianirad, Ali, 49

M
Mohammadi, Hosein, 49
Molaei, Morteza, 109
Mollayosefi, Marziyeh Manafi, 31

N
Nematian, Javad, 31
Nikoukar, Afsaneh, 73

P
Pakrooh, Parisa, 143
Pishbahar, Esmaeil, 31, 91, 109, 125, 143

R
Rashidghalam, Masoomeh, 1, 171
Rasouli, Zahra, 171

S
Sabouni, Mahmoud Sabuhi, 11, 21
Sarajug, Farid Bagheri, 183

T
Torabi, Sasan, 49, 73

M. Rashidghalam (ed.), *The Economics of Agriculture and Natural Resources*,
Perspectives on Development in the Middle East and North Africa (MENA) Region,
https://doi.org/10.1007/978-981-15-5250-2

Subject Index

A

Akaike Information Criterion (AIC), 56, 58, 60, 62, 64, 78, 84, 97, 101
Ali-Mikhail-Hagh (AMH), 96, 101
Analytic Hierarchy Process (AHP), 33, 36
Autoregressive Distributed Lag (ARDL), 146
Autoregressive Integrated Moving Average (ARIMA), 57, 64, 80, 86
Average Treatment Effect (ATE), 4, 91, 97, 104, 106

B

Bayesian Information Criterion (BIC), 56, 58, 60, 62, 64, 78, 84, 97, 101

C

Canonical vine copula (C-vine), 54–56, 58, 59, 61–63, 66
Choice Experiment (CE), 5, 109, 111, 112, 114
Composite Indicator (CI), 32, 33, 35, 40, 44
Cumulative Distribution Function (CDF), 56, 57, 78, 80, 97, 129
Cumulative Rainfall Index (CRI), 57, 58, 65

D

Data Envelopment Analysis (DEA), 6, 12, 13, 17, 33, 183, 186, 188, 189, 193, 197
Decision-Making (DM), 36
Drawable vine copula, 4, 49, 54–56, 60–63, 66, 67

E

Economic Resilience Indicator (ERI), 3, 21–25, 27, 28
Environmental Kuznets Curve (EKC), 145, 162, 164, 165, 167

F

Farlie-Gumbel-Morgenstern (FGM), 95, 101
Food and Agriculture Organization (FAO), 92, 110
Food and Drug Administration (FDA), 24, 26, 28
Full Information Maximum Likelihood (FIML), 97, 105
Fuzzy Analytic Hierarchy Process (FAHP), 3, 31, 35, 36, 38, 42–44

G

Generalized Autoregressive Conditional Heteroskedasticity (GARCH), 6, 171, 175
Generalized Method of Moments (GMM), 152
Greenhouse Gas (GHG), 6, 12, 144, 183–186, 189, 192, 197, 198
Gross Domestic Product (GDP), 1, 6, 110, 143, 155–157, 160, 162–168, 174

I

Integrated Pest Management (IPM), 2, 4, 91–94, 99, 104–106, 198
Intergovernmental Panel on Climate Change (IPCC), 184

M. Rashidghalam (ed.), *The Economics of Agriculture and Natural Resources*,
Perspectives on Development in the Middle East and North Africa (MENA) Region,
https://doi.org/10.1007/978-981-15-5250-2

M
Markov Chain Monte Carlo (MCMC), 56, 153, 154
Modeling to Generate Alternatives (MGA), 18
Multi-Criteria Decision-Making (MCDM), 36

N
Natural Resource Research (NRC), 15, 17, 18

O
Ordinary Least Squares (OLS), 95, 97, 104

P
Pair Copula Constructions (PCC), 53, 54, 56
Partitioning Around Medoids (PAM), 187
Periodic Generalized Autoregressive Conditional Heteroskedasticity (PGARCH), 6, 171, 175, 176
Pesticide-Free Fruits and Vegetables (PFFV), 5, 125, 127, 130–132, 134–137
Principal Components Analysis (PCA), 33, 35
Producer Support Estimates (PSE), 17, 24, 28, 175

R
Regular vine copula (R-vine), 54, 62
Relative Humidity (RH), 51, 57, 58, 65, 74

S
Somatic Cell Count (SCC), 24–28
Stated Preferences (SP), 112
Sustainability Indicator (SI), 2, 11–14, 16–18, 23, 27, 32, 33, 37–41, 44, 45

T
Technique for Order of Preference by Similarity to Ideal Solution (TOPSIS), 33
Temperature-Humidity Index (THI), 4, 73, 74, 76, 78, 80–82, 84, 86, 87
Time-Varying Parameter Vector Autoregression (TVP-VAR), 5, 6, 143, 146, 154–156, 159, 167, 168
Traveling Salesman Problem (TSP), 60
Triangular Fuzzy Numbers (TFNs), 37

V
Vector Autoregressive model (VAR), 173, 175
Vector Error Correction Model (VECM), 175, 176

W
Willingness To Pay (WTP), 4, 5, 91, 93, 94, 96–106, 109–115, 118–120, 127, 128

The manufacturer's authorised representative in the EU is Springer Nature Customer Service Centre GmbH, Europaplatz 3, 69115 Heidelberg, Germany. If you have any concerns regarding our products, please contact ProductSafety@springernature.com

Printed and bound by CPI Group (UK) Ltd, Croydon, CR0 4YY

30/09/2026

02225558-0005